Fiber Optics Communications, Experiments, and Projects

by

Waldo T. Boyd

Howard W. Sams & Co., Inc.
4300 WEST 62ND ST. INDIANAPOLIS, INDIANA 46268 USA

FIRST EDITION
FIRST PRINTING — 1982

International Standard Book Number: 0-672-21834-8
Library of Congress Catalog Card Number: 82-050650

Illustrated by: T.R. Emrick

Printed in the United States of America.

Preface

The transistor transformed the electronics industry in the space of some two decades. It did more than that; it was the germ that seeded a whole new way of life for the world. Bedouins in the desert, peons in the Mexican fields, and peasants in South America and Europe carry transistor radios to their sunup-to-sundown labors, in touch now with the events of the world.

When two catwhiskers were first applied to a block of crystal instead of just one as had been the case to that moment, the potential of the device that came into being was but dimly seen by a few dreamers in the world of vacuum-tube electronics. The first major computer, the ENIAC, was a gigantic apparatus with thousands of vacuum tubes giving off volumes of heat, with a few dedicated scientists and a flock of technicians at its heart. Vacuum tubes would never have spawned today's personal, home, and business microcomputers that are ENIAC's progeny.

Decades ago the dream was for a chicken in every pot, a car in every garage, and a radio in every parlor. Eventually the dream evolved into not only a tv in every living room, but also one in the kitchen, bedroom, and bath of virtually every home in the nation. People everywhere now have access to television via the satellites. But the television receiver is passive; the human animal may sit immobile in fascination a few months or years but becomes restless at last for an active part in the drama. And that need has been provided. The personal home computer provides an active interplay with the screen, and a prediction of a computer in every home is no idle dream. This has come about because of the transistor.

Another such transformation is taking place, barely noticed as yet in the hustle and bustle of everyday happenings in the world of electronics. This change is being thrust upon us by a discovery not of recent date, but of a half century or more ago. Light waves have been with us since the beginning of time, but what has come to pass in recent years is the ability to carry light for great distances through tiny "pipes." Just as we were reaching limits in the crowded electromagnetic frequency spectrum, we rediscovered the fact, almost obvious now, that light is but an extension of that spectrum, capable of carrying hundreds of television channels simultaneously in a slender filament of glass. Once more the vast potential of electronics has been tapped, this time with perhaps even greater significance than when the transistor burst upon the scene. We have a name for it: "fiber optics."

There are a few books available on the physics of light-carrying threads, but they treat the subject deeply and mathematically because they are written for the engineer and physicist. There is no need to add another textbook to the subject at this time. This book is written for the technician and intelligent electronics enthusiast who would like to keep abreast of significant developments in the field, and who would like some "hands on" experience. *Fiber Optics Communications, Experiments, and Projects* opens with a brief overview of the potential light has for the world of electronics, and an introduction into how it is channeled into hair-like threads.

As quickly as possible after covering the major aspects of light and the laws it obeys, we get into some experiments that demonstrate how easy it is to communicate via light beam. Each experiment builds on the one before to convey one more way to work intelligently with light. In the final experiments, you transmit and receive on your own fiber thread.

You begin with a simple light source, a tiny filament bulb known as a "grain of wheat" lamp. A simple photocell receives the light and transforms it into an electrical signal. You progress to the use of a special semiconductor device called a "light-emitting diode," or LED, to transmit light through a fiber, and a simple photodiode to detect that light at the far end of your system.

After the preliminary experiments, you will modulate a light beam with voice or music, and send the information-carrying beam through a fiber to be intercepted by a phototransistor detector. The human voice is changed from waves in the atmosphere to electronic variations in a transistor, to light-wave variations, back again into electronic impulses in the receiver, and finally into sound waves in the air once again.

When the experiments are concluded, you will have become thoroughly fascinated with the potential of this new medium. Your rough experiments are all open and very interesting, but now you want to build a few "smooth" projects of lasting and practical value. In their turn, these projects are but one step toward the tomorrow that is certain to come for fiber optics. With the experiments and projects behind you, you will be ready to grasp the opportunities that are already arising in the industry for putting fiber optics to practical use.

Some of the "tomorrows" are already "todays." There is an office copier on the market today that owes its almost unbelievable contrast and sharpness in copying on ordinary paper to fiber optics. As for opportunities, home computers have two most annoying problems despite their popularity: they suffer from static electricity, and they give off radio-frequency interference. One of the most promising cures for these problems lies in fiber optics. It all begins here. Where it ends is up to you, because once you have grasped the principles of working with light, you will go on to solve such problems as these.

A number of suppliers of fibers, couplings, phototransistors, and LEDs have been most courteous in supplying materials from which to develop the experiments and projects herein. These are listed, with addresses, in Appendix D. In particular, I thank Corning for a generous supply of fiber, and AMP Incorporated for use of their engineering kit and for their kind permission to reproduce a number of illustrations from their excellent introductory handbook. Sylvania and General Electric supplied sample phototransistors, and Belden provided some multiple-strand cable. Radio Shack was most helpful in supplying hardware and electronics, and their part numbers are included for reference in obtaining components with which to duplicate the experiments; the "Archer" trademark designates the Radio Shack product.

The following people helped by providing technical materials and information as well as advice and assistance in writing this book: Peter Von Keyserling of AMP Inc., Hank Schroeder of Motorola, William

Hegberg of Opto Technology, Inc., Bill Nunely of TRW Optron, and James Pletcher of AMP Inc. Thanks are also due to Math Associates and Motorola for their assistance. The support of these people and companies is greatly appreciated.

I wish also to thank Richard Cunningham, editor of *Electro-Optical Systems Design* magazine, for his helpful encouragement in undertaking the task of writing this book. And I owe a debt of gratitude for invaluable advice and assistance to Dr. Jonathan Titus of The Blacksburg Group. His guidance has been one of those remarkable balances between praise and stern criticism which reveal the born teacher.

May this be but the first of your fascinating explorations into the most significant development since the transistor!

WALDO T. BOYD

To My Grandson
Duane Brown

" . . . If comparison be made with the sum total of all former human achievements it will be found that the discoveries, scientific advancement and material civilization of this present century have equalled, yea far exceeded, the progress and outcome of one hundred former centuries."

—Abdul Baha' (1844–1921)

Contents

Section 1: Fundamentals of Fiber Optics

Section 2: Experiments

Chapter 7

Chapter 8

Chapter 9

Chapter 10

Chapter 11

Chapter 12

Chapter 13

Chapter 14

Section 3: Projects

Chapter 15

Chapter 16

Chapter 17

Chapter 18

Chapter 19

Section 4: Appendixes

Appendix A

Appendix B

Appendix C

Appendix D

Appendix E

Appendix F

Appendix G

Appendix H

Appendix I

Appendix J

Appendix K

Section 1
Fundamentals of Fiber Optics

1. INTRODUCTION TO FIBER OPTICS
2. BASIC PRINCIPLES OF FIBER OPTICS
3. WHERE IT ALL STARTED
4. THE "NUTS AND BOLTS"
5. BASIC PRINCIPLES OF GLASS FIBERS
6. SIMPLE SYSTEMS

1 Introduction to Fiber Optics

Our universe is energy; everything we sense, measure, and experience is energy in one form or another. Sometimes energy is obviously in motion; sometimes it seems to be at rest, ready to perform some task at Nature's behest or man's command. The more we contemplate energy, the more awe-inspiring it becomes. All-pervasive, in and around us, governing our every action and reaction, determining our life or death, our coming and going, energy is the stuff of life and of the cosmos of which we are materially a part.

This is a book about energy. The particular kind of energy we shall consider is electromagnetic in nature, similar to the radio waves that bring our transistor radio to life with the music of the latest combo. However, the energy we shall explore is in very, *very* rapid vibration as compared to the broadcast radio wave. There is a name for energy that is alternating this rapidly: *light*.

Light was not often thought about in terms of energy before the invention of the laser. Now we know it as one of the most potent forms of energy we can imagine.

Few of us have taken time to think about the structure and function of our eyes. The eyes we take for granted during every waking moment are tiny, highly efficient "television receivers" that can differentiate among a multitude of frequencies and let us see color. Each specific color represents a certain frequency of electromagnetic energy, just as does one station tuned in on our radio. The retinas of our eyes contain marvelous "antennas" and detectors which respond to these electromagnetic waves of energy in a manner that is, even in this enlightened scientific age, a profound mystery.

Our eyes are blind to the radio frequencies that our transistor radio is capable of sensing, and the radio is "deaf" to the frequencies our eyes can see. Our eyes are likewise blind to gamma rays that are emitted from radioactive substances or the physicist's cyclotron. Our marvelous light sensors gradually lose their ability to sense as the frequency of light extends down into the infrared region or up into the ultraviolet region.

Only a few years ago, the terms megahertz, gigahertz, and femtohertz were unknown, as were the facts involved in this very-high frequency portion of the spectrum. A few pioneering physicists wondered if we would one day be able to generate light waves in much the same manner that we do radio waves, up, up, and yet farther up into the extremely short waves measured in angstrom units (10^{-8} cm).

We reached that level of understanding and in due course were able to generate light directly, rather than by such means as a hot length of tungsten wire. Today our little hand-held calculators wink in response to our pressing the keys; the winking red lights originate in solid-state light generators called *light-emitting diodes* (LEDs). And in another way we can now cover the entire visible light spectrum with a device we call a *tunable laser.* ("Laser" is an acronym for *L*ight *A*mplification by *S*timulated *E*mission of *R*adiation.)

As we rolled forward the upper limits of generating electromagnetic waves, the amount of information we could code, or modulate, into a given place in the spectrum increased geometrically. Early in the days of radio communication, the cumbersome Alexanderson alternator served as the generator of electromagnetic waves at about 10,000 cycles per second. (In later years "cycles per second" came to be known as *hertz* (Hz), in recognition of the early experimenter who first identified radio waves, Heinrich Hertz.) The ten-ton alternator later was replaced by a water-cooled vacuum tube light enough to be hefted in one hand.

The amount of information that can be conveyed from one point to another with a carrier wave of only 10,000 Hz is limited to but a portion of the human voice range. When the vacuum tube was developed, the frequency of the radio carrier wave could be increased because it was no longer limited by the inertia and tensile strength of rotating machinery. With every doubling in frequency, the amount of information that could be conveyed from one point to another was increased fourfold.

And then came television. Television demanded a bandwidth that in its early days was considered "impossible" when thinking in terms of the radio-broadcast frequency spectrum. The video intelligence—that is, the picture elements—required nearly four megahertz (four million hertz) of the known useful electromagnetic spectrum.

This incredible demand for a "lion's share" of the spectrum came at a time when we had only barely exceeded this figure at the top of our short-wave bands, pioneered by the amateur radio fraternity. But the limits were extended again and again in home workshops and in laboratories, until at last direct generation of light became possible in the same way that radio waves of lower frequencies are generated today, by solid-state oscillators.

The ability to modulate, or impress intelligence upon, that electromagnetic energy kept pace. Voice communication on a beam of light, although known decades earlier in a gross sense, became possible and then practical, and now it is becoming almost commonplace in a refined sense.

The range of man's creative imagination far exceeds his ability to perform immediately. Talking on a beam of light had been done experimentally even before the turn of this century. In those early days of opening the door to radio communication, the empirical method (a word that roughly translates to "cut and try") succeeded even though the reasons why it worked were obscure.

Even today, talking on a beam of light arouses in the experimenter a fascination that has to be felt to be understood. Other forms of electromagnetic communication are sensed only indirectly; light can be seen directly, although sensing its modulation is limited to about a 16-Hz signal. This is a very low figure, relatively speaking.

Anyone interested in communicating with a light beam should at some time perform the experiment personally, even though he might feel he has long since left these simple things behind. A light bulb, a photoelectric cell, an amplifier, and a microphone are all that are necessary to talk across a room even in daylight, and across hundreds of meters on a clear night. With a lens or two, nighttime communication can be carried on for many kilometers under ideal conditions, mountaintop to mountaintop. The atmosphere, however, is but one of many media through which the electromagnetic energy called light will propagate.

While fascinating when first performed, the experiment will soon convince those participating that the atmosphere as a dependable medium is limited in many ways. The beam is attenuated by smoke, fog, or haze. It can be intercepted easily by someone other than the intended party. It scatters from the point of origination, increasingly for the entire distance of its travel. Ambient noise (extraneous bursts of light) interferes with communication, as lights from any background source will register in the receiver photoelectric cell.

In the early days of radio, experimenters sometimes demonstrated what was then known as "wired wireless." Radio waves that were ordinarily sent out into the atmosphere from an antenna could instead be directed along a wire with minimal radiation into surrounding space. In a commercial adaptation of this principle, wired-wireless intercom units for homes and businesses may be purchased at most electronic supply houses. These use the electric power wiring as conduits for communicating, by placing the intelligence in a modulated carrier wave that passes through a building on the wiring used for the 60-Hz power.

In transmitting current along a wire, ordinary power frequencies (60 Hz in the USA) make use of the total cubic content of the conductor. As frequencies are increased into the kilohertz range, the energy tends to travel increasingly near the surface of the conductor. From about 10 kHz upward, the effect becomes pronounced, until in the 50-megahertz (MHz) range and upward it is common practice to use hollow conductors in higher-power equipment to eliminate the waste metal, and to reduce the eddy-current heating effect within the core metal of a solid conductor.

When multiple wires are cabled, with each pair carrying independent information such as in a telephone cable, there is an effect called *cross talk* that becomes more pronounced the higher the frequencies involved. Capacitive and electromagnetic coupling occur, causing a measurable transfer of energy from one pair to all other pairs in the cable. While twisting each pair independently of other pairs does reduce the cross talk somewhat, enough is left to be of concern. Thus, there is a practical limit to the amount of data that can be carried in a copper-wire cable.

Experimenters working with ultrahigh-frequency (uhf) radio waves developed what are called *waveguides* to conduct radio-frequency energy from point to point more efficiently than through wires. A waveguide is little more than a rectangular pipe when used for uhf energy. Such pipes are referred to as "plumbing" by those who work with microwaves.

Waveguides are usually made of copper and brass, plated inside with gold or silver and highly polished, the better to conduct at the high frequencies involved. The inside dimensions—length, width, and depth—are very carefully calculated to accommodate one particular frequency. Early waveguides were huge, cumbersome affairs larger than the foundation timbers of a house. As frequencies increased, the dimensions of waveguides decreased, until they became so small that when any appreciable power was employed, the waveguide arced across inside, and the high polish was destroyed.

Light energy will travel through these waveguides, although not very efficiently. The light bounces off the inner walls at all angles because the dimensions are too great for the very short wavelengths, and very little of the original light energy reaches the far end of even a relatively short length of waveguide, particularly if there are any bends involved. By reducing the dimensions of the waveguide, the angles of light incidence and reflection become smaller, and an improvement in propagation efficiency is noted.

Instead of air as a medium of propagation for light, we can prepare a glass rod by cutting it at right angles to its length and polishing the ends. When we place a light source at one end, light will be seen at the other. The rod may be heated and bent into a "U" shape, and light will travel from one end to the other. Some light will escape all along the rod length, so efficiency must be improved if we are to enjoy transmission over significant distances.

A glass or plastic tube filled with water will serve nicely as an experimental "light pipe" to demonstrate the principle involved. A glass pipe is rigid and must be preformed into the desired path. We can fill a long, transparent, flexible plastic pipe with water, alcohol, glycerine, or other transparent fluid and place a light at one end and a photocell at the other end. Such a contrivance will work surprisingly well as a medium of communication, as your experiments to follow will show.

While this provides an admirable demonstration of the principle of light conduction through other than air or vacuum, and can even do an adequate job under limited practical circumstances, a more versatile light conductor is needed. Ingenious glass-making engineers formed a light conduit by stretching a glass rod under heat until a tiny glass fiber was formed. This fiber carried light for relatively short distances, but it was an encouraging beginning in technology. Some of the energy escaped along the surface of the glass thread, and some was absorbed into the medium itself in the form of heat.

Glass is made from sand, which is silica. By making glass fibers of ultrapure silica and doping with special formulations, and by cladding the surface with oils, plastics, and glass of different compositions, glass threads or fibers were developed to carry light quite well for distances of 300 meters and beyond. The search for increased efficiency in light conduction continued, and recently a fiber has been developed that does a more acceptable job of light transmission over many kilometers from point to point with little loss.

Westinghouse has recently conducted an experimental space manufacturing process in a successful attempt to produce super-pure glass. Scientists have debated for some years whether the bubbles that arise in glass making might be eliminated in a "zero-g" environment. This experiment, conducted 120 miles high at the peak of a rocket trajectory, proved that the bubbles would migrate toward the lower temperature glass surface, leaving the glass ultra pure. This purity is vital to light-carrying fibers.

Where a copper wire in communication service would be subject to electromagnetic interference, glass fiber is virtually immune. Where a wire can be "tapped" (surreptitiously compromised by an eavesdropper) quite easily, a glass fiber can be tapped without detection only with the greatest difficulty. While a reel of copper wire can weigh several tons, its glass counterpart weighs only a few hundred pounds, and most of that is the reel itself. A few ounces of sand become a kilometer of fiber waveguide.

Glass is but one of a number of mediums through which light will propagate. Today, plastic fibers are playing a part in the growth of what has become a multibillion-dollar industry. Guiding light through a strand of glass is called *fiber optics.* Included in this term are the amplifiers, modulators, connectors, and other accessories required to complete a communications system.

2 Basic Principles of Fiber Optics

We may wish to send light through a fiber with no intelligence on it whatever, for purposes such as conveying a strong spot of light into the stomach. Another fiber picks up the reflected light, and the doctor inspects the stomach lining for ulcers or indication of cancer. For this application, we need a light source such as a light bulb, a dual fiber cable, and a coating that makes the fiber practical to handle. Such a cable is called a *bifurcated bundle* because of its dual nature.

A prime example of a practical application for a coherent fiber-optic bundle has been developed by John M. Franke and David B. Rhodes of the Langley Research Center of NASA (Figs. 2-1 and 2-2). Physicians often have need to see into the internal cavities of living humans, and veterinarians likewise into the animals they care for. Direct vision into such areas as the colon, the esophagus, and the stomach is extremely difficult without the aid of instruments such as the endoscope.

The *endoscope* is a fiber-optic device that sends a beam of light via fiber into the cavity being inspected and carries the reflection of that light back into a viewing lens. A coherent bundle of fibers affords a direct image of the inner surface of the cavity. However, such a view is somewhat circumscribed, especially when teaching a class is involved. Only one student at a time may look at the object in question.

By coupling the endoscope to a television camera via another fiber-optic bundle, with matching lenses, the *arthroscope* is born. With it, the inner surfaces of the body cavity under examination can be seen on a television screen, in color, and at a size readily visible to a

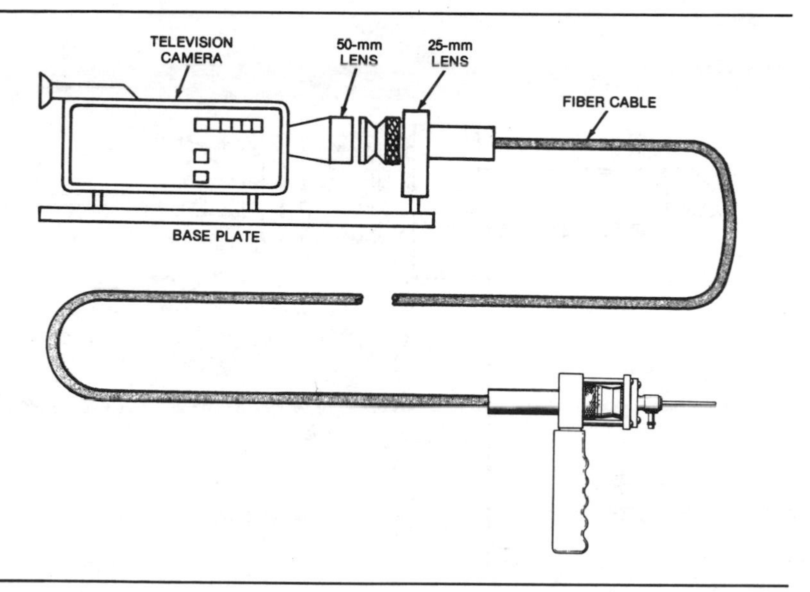

Fig. 2-1. Fiber-optic coupler used to link arthroscope with color television camera. (Courtesy NASA)

number of students. Several other advantages accrue, among them the ability to keep the television camera several feet away from the surgical procedure. The fiber cable in the unit shown is 3 meters (about 10 feet) in length. It could just as easily have been 300 meters long.

Single fibers can be cabled into a bundle to increase total light conduction, as in Fig. 2-3. Except for ultimate cost, there is no upper limit to the number of fibers that can be bound. The bundle may be round along its length for lacing convenience and the ends brought into a square or a rectangle for viewing. By highly polishing one end of the bundle to accept light energy more readily for transmission, and making the other end translucent to act as a viewing screen for images, a total image may be transmitted without scanning.

If we wish to project an image into one end of the fiber bundle and receive it intact and oriented identically to the original, the fibers in the "receiving" end must be in inverse order along the horizontal dimension; otherwise the received image will be presented right to left rather than left to right as transmitted. We would in effect be

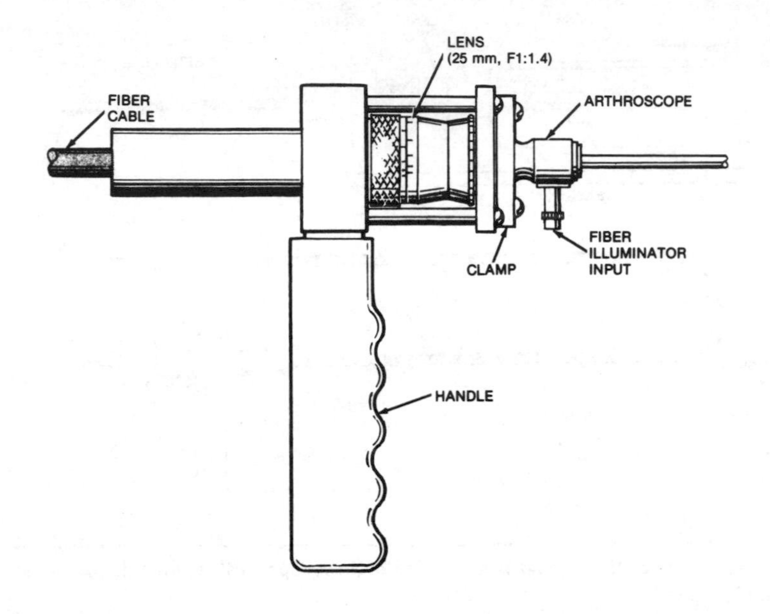

Fig. 2-2. Hand-held portion of coupler in Fig. 2-1. (Courtesy NASA)

looking at the back of the image, rather than the front. With a large enough bundle, we could construct a "mirror" by bending the cable into a "U" shape, standing in front of one end and viewing our own

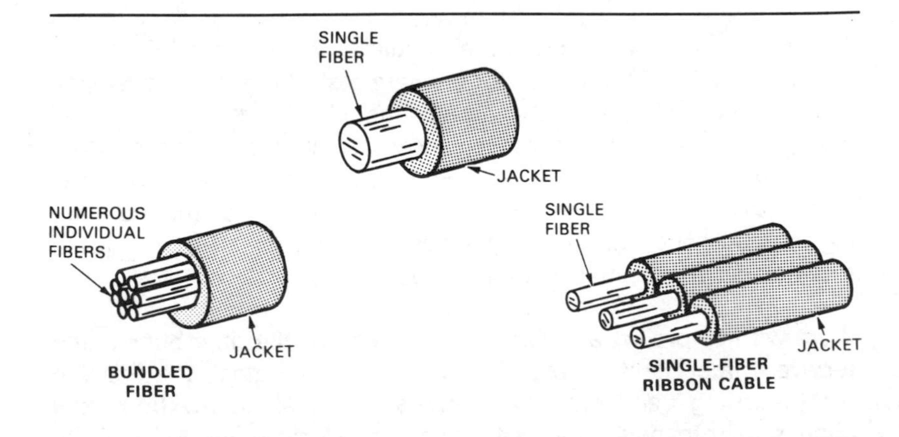

Fig. 2-3. Fiber-optic bundles and single fibers. (Courtesy AMP Inc.)

image at the other, as in Fig. 2-4A. It would be a very expensive mirror, of course, and economically impractical, but there are applications of fiber optics that employ a principle quite similar to this. An example is a means of enhancing the output of a cathode-ray tube.

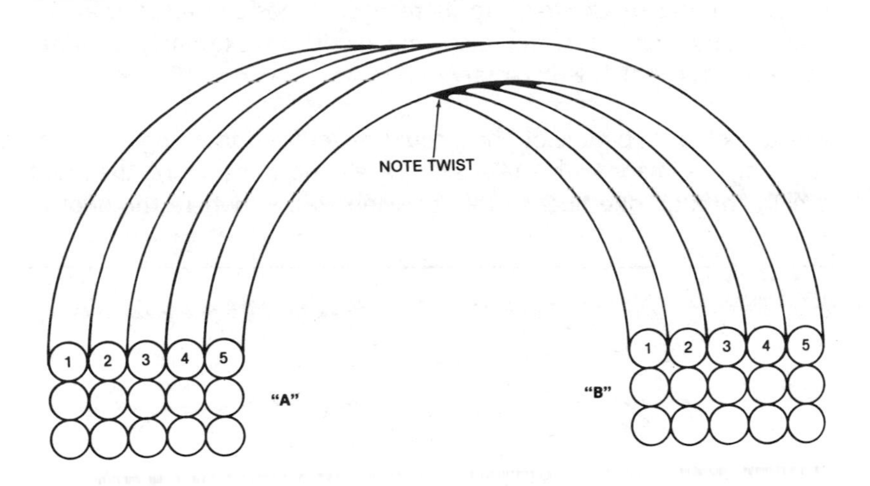

(A) Fibers transposed.

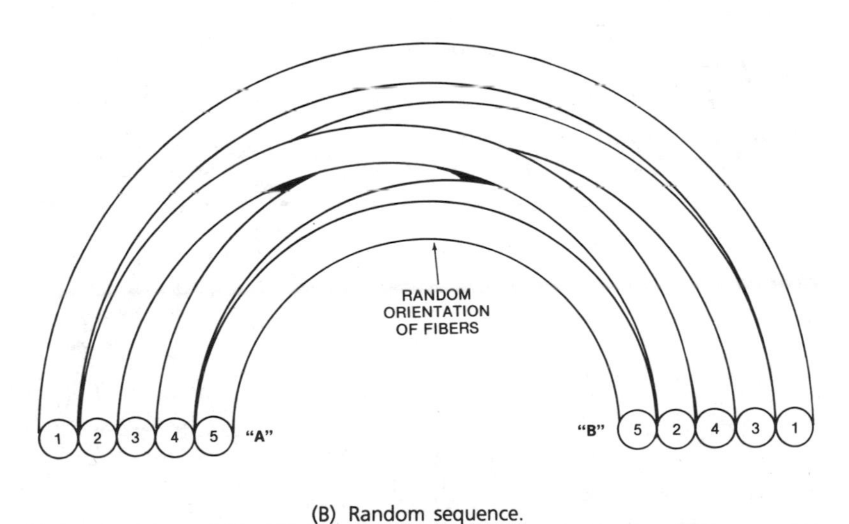

(B) Random sequence.

Fig. 2-4. Transmission of images by fiber-optic cables.

If the individual fibers at the receiving end are arranged in a random sequence or in a systematic sequence different from the order of fixation in the transmitting end, an incomprehensible pattern of light and shadow will appear at the receiving end (Fig. 2-4B). Reconstruction of the original image requires only a short bundle of fibers inserted between the received image and the viewer (Fig. 2-5). The inserted bundle reroutes each strand to its respective position as seen at the transmitting end—a secret code machine! An extremely accurate fiber-for-fiber match is necessary.

A large coherent bundle of fibers could be reduced to a single fiber by scanning the image with a television "eye" and transmitting the result serially through one fiber to the receiving station, where the original

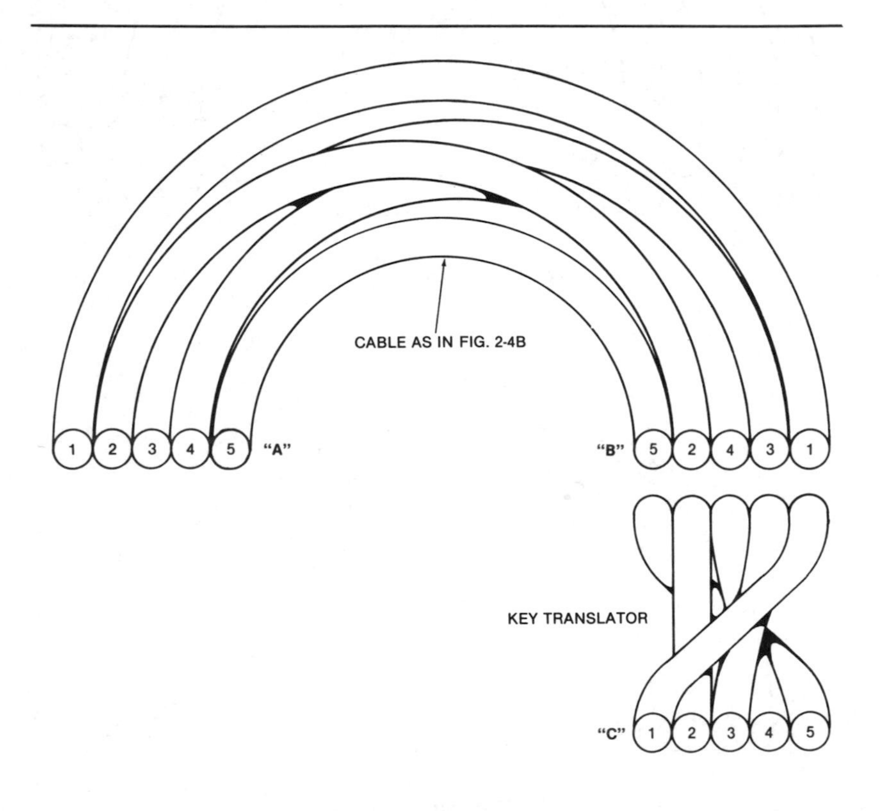

Fig. 2-5. Use of translating coupler.

can be made to appear on the raster of a picture tube. The received picture could be rendered incomprehensible by sending the data in a different order than that seen by the scanner, to be rearranged by a short "key" fiber-optic bundle placed in front of the picture tube to reroute the image elements to their original positions.

3 Where It All Started

The name "fiber optics" is credited to N. S. Kapany, who coined the term in his book by the same name in 1956. He defines fiber optics as "the art of the active and passive guidance of light (rays and waveguide modes) in the ultraviolet, visible, and infrared regions of the spectrum, along transparent fibers through predetermined paths." He is also credited with the invention of coated glass fiber.

The Development of Fiber Optics

Without coining a special name, John Tyndall gave a demonstration of the light-pipe phenomenon in 1870 at the Royal Society in England. He illuminated a container of water from the inside and opened a hole in its side. Light was conducted with evident success along the curved path of the flowing water (Fig. 3-1).

In 1880, Alexander Graham Bell proposed that speech could be transmitted through or on a beam of light. He referred to his device as a *photophone*.

British Patent Spec 20,969/27 was registered to J. L. Baird, and US Patent 1,751,584 was granted to C. W. Hansell in 1930 for scanning and transmission of a television image via fibers. Also in 1930, H. Lamm, in Germany, demonstrated light transmission through fibers. The next reported activity in this field took place in 1951, when A. C. S. van Heel in Holland and H. H. Hopkins and N. S. Kapany investigated light transmission through bundles of fibers. While A. van Heel

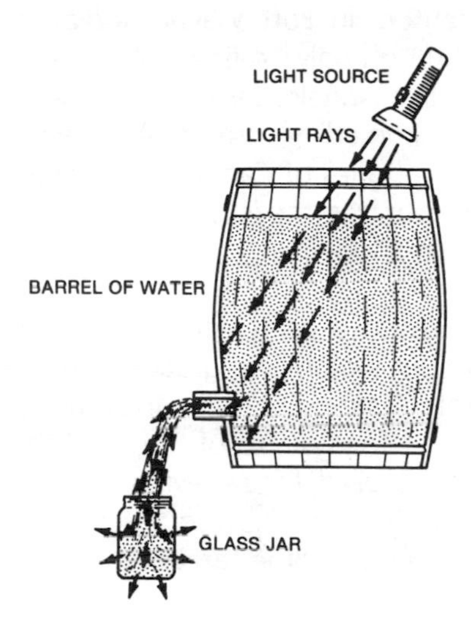

Fig. 3-1. John Tyndall's demonstration (modified with modern flashlight).

coated his fibers with plastic, Kapany explored fiber alignment, and as reported in his book, *Fiber Optics*, produced "the first undistorted image through an aligned bundle of . . . uncoated glass fibers."

All was quiet in the new science until 1967, when K. C. Kao and G. A. Bockham of England's Standard Telecommunications Labs suggested that a new communications medium using cladded fiber might be in the offing. From then until the present the technology has grown enormously, at first quite slowly, but during the 1970s almost geometrically.

At this writing, government and defense agencies, telephone companies, and a widening range of other private companies are turning to fiber optics for telecommunications and other uses, many under the wraps of military and proprietary secrecy. Each new user is attracted by the nature of optical transmission that does not simply reduce but avoids entirely the risks of short circuits, wire-tapping, and electromagnetic cross talk. American Telephone and Telegraph is preparing a Boston-Washington fiber-optics link of over 600 miles, to carry 80,000 simultaneous telephone calls through 19 all-digital switching substations in seven states plus the District of Columbia. The installation is expected to have reduced construction and operating costs some $50 million by 1990.

General Telephone of Indiana has a 3-mile link between two switching centers in Fort Wayne, Indiana; it carries 5000 telephone circuits through only 14 fibers. A similar system was scheduled to be installed in San Angelo, Texas before the end of 1981. Bell of Canada has plans for immediate use of fiber technology, as does Alberta Government Telephones of Canada. The list is growing as company after company recognizes the superiority of fiber optics over the old copper-wire technology.

England expects completion of nearly 450 km of optical-fiber cable under water. Two grades of fibers are to be used, one for higher-bit-rate systems of 140 Mbps, using laser injection techniques, and another for systems of 8 to 34 Mbps, using LEDs. Avalanche diodes are to be used for receiving. France and Germany and other European countries are likewise entering the field, primarily in telephone work, but also in other areas.

By 1990, AT&T will be laying a fiber cable that will span the Atlantic ocean. Cables between other countries are expected to follow in rapid succession. Electric utilities are installing fiber cable links for communications with substations; these replace both the venerable carrier-telephone equipment that multiplexes the power lines and over-the-air transmissions. Fiber cables are also being laid underground as security links (Fig. 3-2). Television satellite links are making use of fiber optics as well (Fig. 3-3).

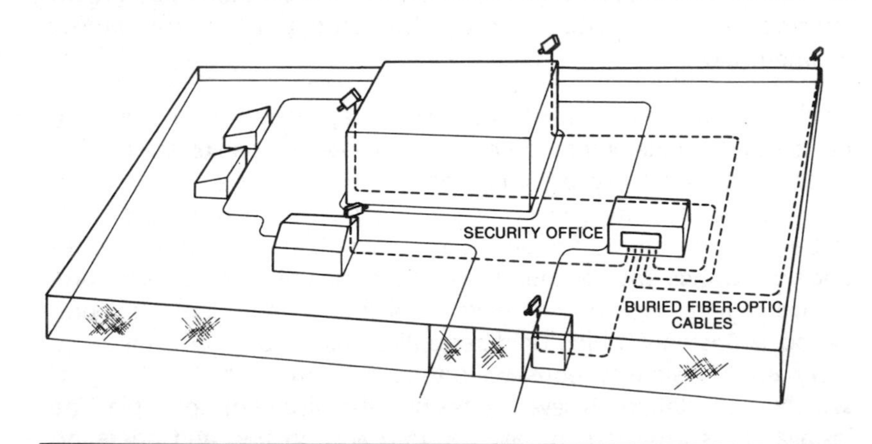

Fig. 3-2. Use of fiber optics in an industrial plant.

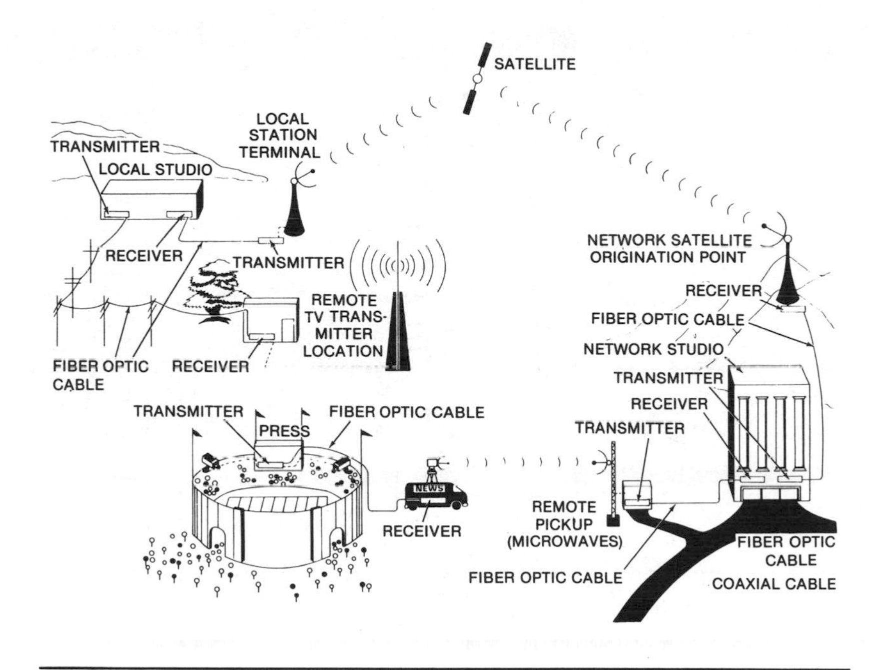

Fig. 3-3. Use of fiber optics in broadcasting. (Courtesy Valtec, a Philips-M/A-COM Venture)

The Usefulness of Fiber Optics

The electromagnetic spectrum lies from the "subsonic" range of a few hertz to cosmic rays at 10^{22} Hz (Fig. 3-4). Within this extensive span of frequencies, the region of interest in fiber optics lies near 10^{14} Hz, from infrared through the visible spectrum to ultraviolet.

Modulating — that is, impressing information upon — any selected single frequency can take a number of forms, three of which are illustrated in Fig. 3-5. Further, communications may be "multiplexed" in a number of ways, extending the potential usefulness of fibers in short-run installations. Formerly, it was common to talk about how many hundreds of channels could be packed into one fiber, but there are also many uses for fibers that will carry only a few channels. Wavelength division multiplexing, wdm, provides a means to include several subscriber lines within a common fiber along the trunk line,

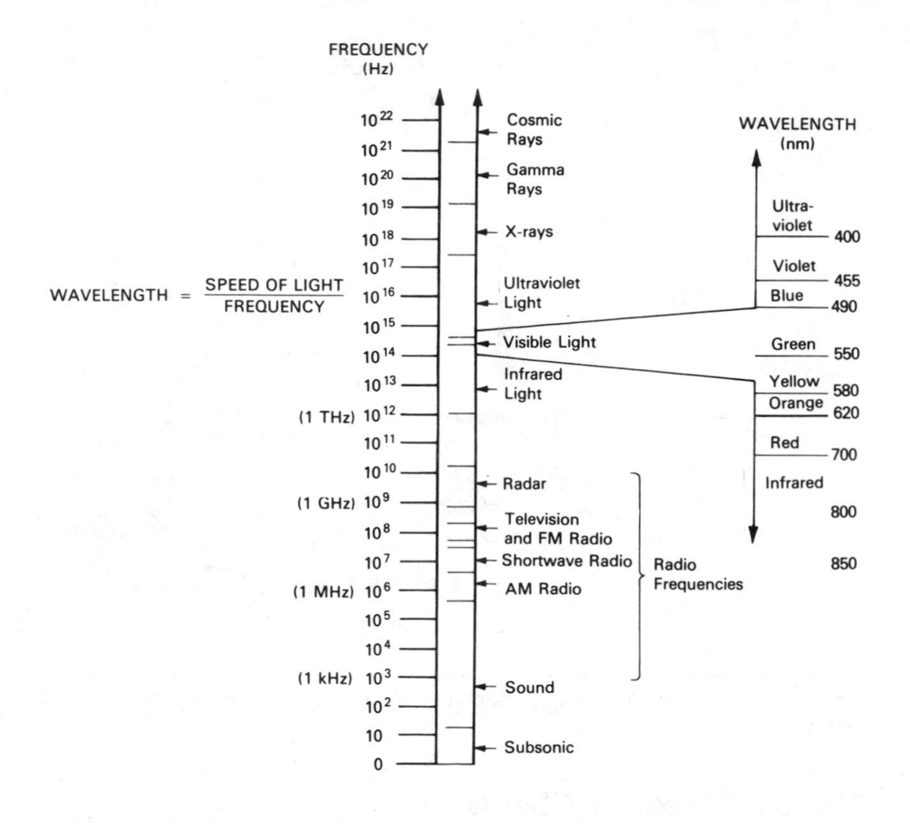

Fig. 3-4. The electromagnetic spectrum. (Courtesy AMP Inc.)

and to separate them at the far ends. This system functions by combining and dividing signals according to their wavelengths. As an analogy, we may think of the prism separating a ray of sunlight into its spectrum of colors (Fig. 3-6). Each color represents a subscriber's private communications line.

While comparison of fiber-optics cables to conventional coaxial cables is startling, leaving the latter so far behind as to appear almost unbelievable with regard to capacity for communications transmission, system costs cannot be compared quite so readily. Because the fiber-optics cable must have more mechanical protection, it may cost more to produce. Because fiber-optics systems are at present just barely out of the laboratory, whereas coaxial-cable systems have been commercially available for at least 40 years, fiber-optics systems are generally more expensive.

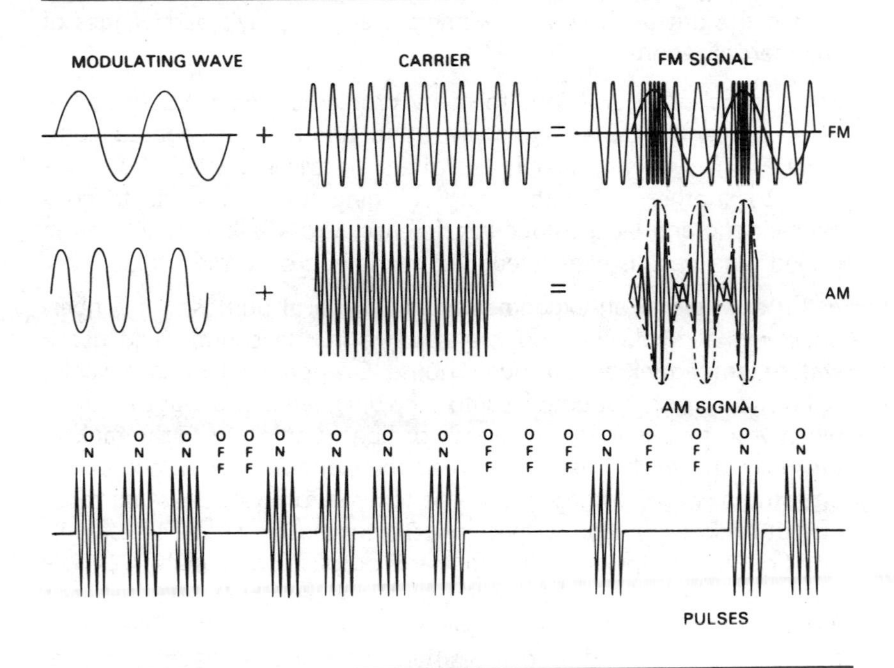

Fig. 3-5. Carrier waves and modulation signals. (Courtesy AMP Inc.)

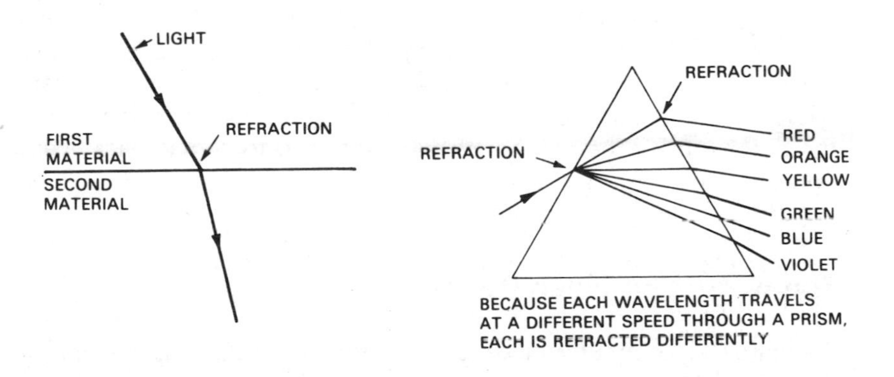

Fig. 3-6. Refraction of light. (Courtesy AMP Inc.)

When full use is made of the wideband capabilities of fiber-optics systems, and when allowance is made for the decreased need for booster amplifiers as compared with coaxial systems, fiber-optics systems come out unchallenged winners even in today's early stages of commercialization.

Fiber optics offers solutions to many problems long encountered in wire, concentric-cable, microwave, and antenna-propagated communications systems. Fiber optics offers electrical isolation from one point to another. It is immune to lightning discharges and to electromagnetic and electrostatic interference. Cross talk encountered in cabled wire pairs is eliminated. Security is improved manyfold.

As an example of an experimental and practical purpose for a fiber-optics signaling device, suppose you live in the country and use a water-storage tank at your pump house. Or, perhaps you have a solar still on your roof that empties into a holding tank. You want to know when your tank is full. Two strands of fiber running from the tank to the point at which you wish to read the signal will work admirably; one strand is for light input, and the other is to conduct the mirrored output back to the readout position. Or, perhaps you want to use only one fiber strand for both the input and output light signals. A simple system of mirrors in the tank will reflect light back into the fiber when water is present and shunt incoming light away from the fiber when no water is present, due to the difference in indices of refraction of water and air.

The weight saving inherent in fiber cables is highly important in aircraft and spacecraft (for increased payload of the latter in particular). Although of lesser import for ground installations, weight saving does become significant when large spools of cable are transported. With weight saving comes saving in overall costs. The amount of signal power required for a given point-to-point transmission is less, and there is, for these reasons, an overall cost saving to be had with fiber optics in most large-scale installations.

The Nature of Fiber Optics

Standard Telecommunications Laboratories in England is generally considered to have been the first to advocate glass fibers for long-distance communications and information exchange. Even though typical losses in the existing fibers in 1968 amounted to no less than 1000 decibels per kilometer (dB/km), the workers at the British labs were confident that development of refined fiber material would lower these losses to an acceptable level. Two years later, Corning Glass Works in the United States achieved a reduction of losses to what was then considered to be a remarkable low of 20 dB/km for single-mode fibers.

Losses

Today, single-mode fibers are available with losses as little as 0.5 dB/km, and Corning's Code 1505 and 1516 fibers run 2.9 dB/km at costs competitive with copper wire. With fibers of this purity, kilometer runs without booster amplification are commonplace, and runs of up to 10 km are possible before relay amplification is mandatory. Bandwidths of 39 MHz at a wavelength of 850 nanometers (nm) are claimed.

The prime cause for high attenuation (loss, roughly equivalent to resistance in copper wire) in the 1968 fibers was metallic impurities in the glass, although water, strange as it may seem, also played a part in obstructing the signal. Infrared absorption in the fiber now establishes the lower limits of attenuation at longer wavelengths. It is to the advantage of virtually all users of fiber optics to attain the lowest attenuation possible within the range of practical economy. Repeater amplifiers will permit transmission for virtually around-the-world reception, but each repeater creates a potential failure point in the system, as well as an increase in cost. Thus, it is desirable to minimize attenuation to keep down the number of repeaters required.

Propagation Time

If we keep in mind that light does not travel instantaneously, whether in a vacuum or through any solid medium, we can see that two light rays traveling different distances will arrive at different times. While this fact is quite basic, the concept of light "taking time" is not often considered in the ordinary work of technology. Light is so speedy—that is, it takes such a minute fraction of a second to travel one kilometer—that it is ordinarily considered to take no time at all for practical purposes. However, in fiber optics as in radar work, the transit time of light through some media becomes quite important, indeed.

Not all rays entering a fiber at its polished end will enter exactly parallel to the fiber axis. And even when a light ray does enter exactly at and parallel to the axis, it will tend to travel in a straight line when it encounters a curve in the fiber. Thus, *within the aperture of acceptance of the fiber*, rays will be entering at all possible angles. They will be reflected at the surface (the boundary between two different indices of refraction) and will follow a zig-zag course for the entire length of the fiber.

In an absolutely straight length of fiber, one ray may travel through the axis and may reflect but once, in the exact center of the length, and emerge from the other end at the axis point. Another ray may reflect from side to side many times before emerging (Fig. 3-7). The greater the number of reflections, the longer light will take to arrive at its destination. Between the two extremes, rays experiencing varying angles of reflection or refraction will arrive at different times. Any modulation existing on the basic light wave will be distorted to the extent that the light rays follow different paths from entrance to exit.

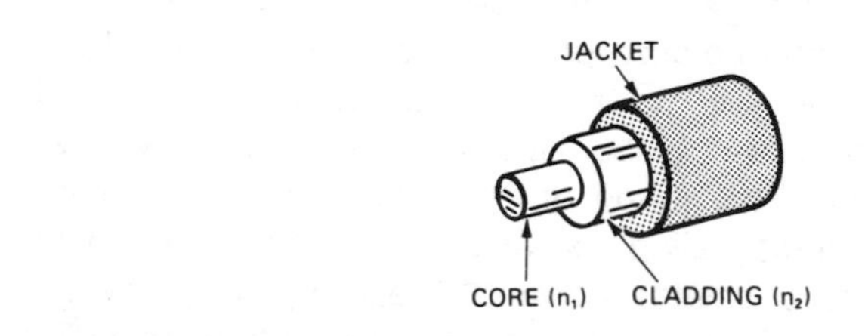

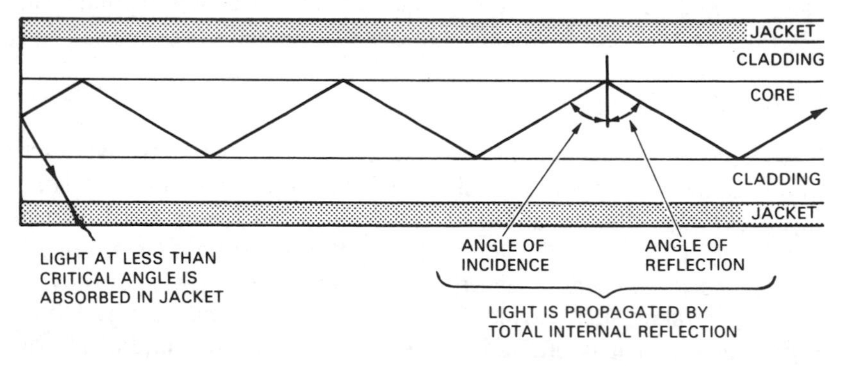

Fig. 3-7. Reflection and/or refraction of light at boundary of core and cladding of an optical fiber. (Courtesy AMP Inc.)

Difference in propagation times is referred to as *modal dispersion* in fibers. Such dispersion limits the bandwidth of some fibers to as little as 20 MHz in a kilometer. For many applications, a bandwidth of up to 300 or 400 MHz is needed. Thus, there was incentive to develop a fiber which would reduce such dispersion to a minimum.

Through ingenious manufacturing techniques, a *graded index* fiber was developed (Fig. 3-8). By causing the light to travel faster the farther away from the axis the ray wanders, the differences in travel time of the rays are minimized. (Fig. 3-9 compares the refractive-index profile of a graded-index fiber with that of a step-index fiber.) Graded-index fibers can propagate bandwidths as high as 600 MHz in a kilometer length. Compare this to the standard radio broadcast band, which extends from 0.54 MHz to 1.6 MHz. We could operate 600 radio-broadcast spectrums simultaneously through such a fiber, with room for 600 times the number of broadcast stations presently accommodated. At 6 MHz each, we could operate 100 television channels simultaneously. This on *one single strand* of fiber, and a bundle no larger than your finger might contain a thousand strands.

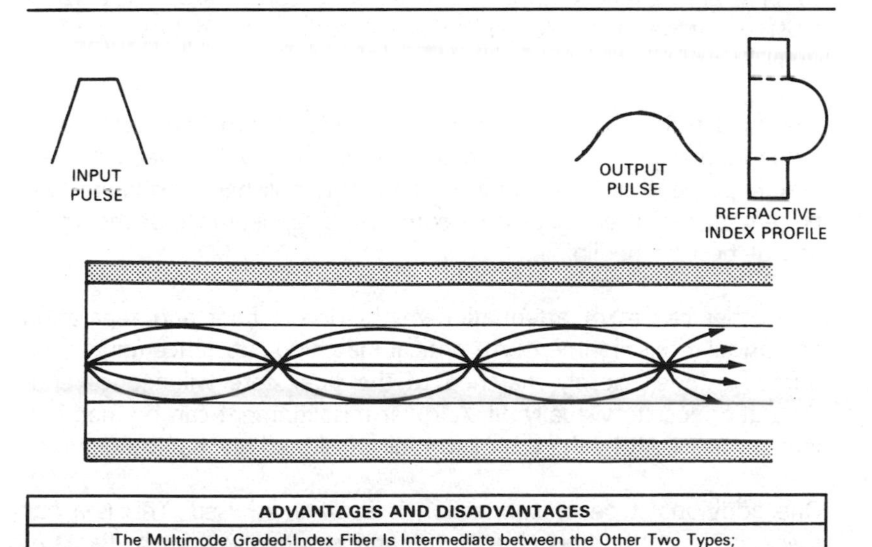

ADVANTAGES AND DISADVANTAGES
The Multimode Graded-Index Fiber Is Intermediate between the Other Two Types; Its Advantages and Disadvantages Lie between the Other Two.

THE FIBER IS SUITED FOR MEDIUM DISTANCES AND MEDIUM OPERATING SPEEDS.

Fig. 3-8. Multimode graded-index fiber. (Courtesy AMP Inc.)

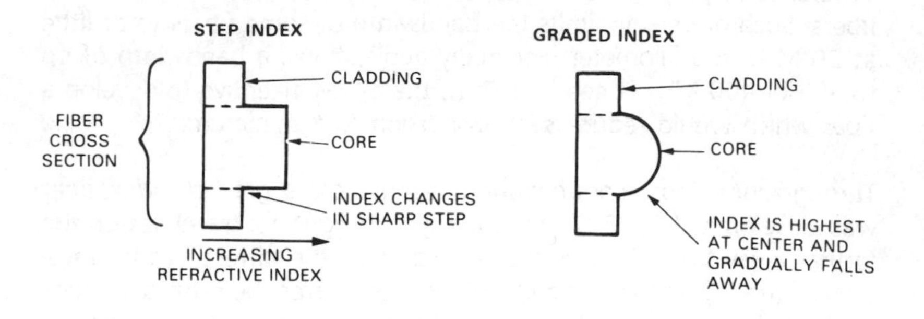

Fig. 3-9. Variation of refractive index in step-index and graded-index fibers. (Courtesy AMP Inc.)

Coupling of Fibers

Ideally, a fiber would be one continuous strand from input to output. In practice, fibers are manufactured in discrete lengths of up to about a kilometer, although longer runs are available. If a fiber must be coupled end to end with a second fiber to continue a run, a splice must be made, and each splice will introduce some loss, however small it may be.

The most serious loss will occur if the ends of two fibers are axially misaligned (Fig. 3-10). Assuming that each end has been broken exactly perpendicular to the axis and highly polished, this lateral displacement can at worst cause a complete disappearance of the signal and at best be negligible.

Two other causes of attenuation might occur: fiber end separation and angular misalignment. An optical matching substance (glue) with refractive index closely matched to the fiber core will reduce end-separation loss to virtually nil. Angular misalignment can be made almost nonexistent by the use of specially designed mechanical fixtures.

One additional type of connection is sometimes used. This is a *hot splice*, in which the fibers are butted and welded together by flame or electric arc.

Refer to the three loss-mechanism graphs in Fig. 3-10. It is evident that three means of modulating the light beam are suggested. Analog

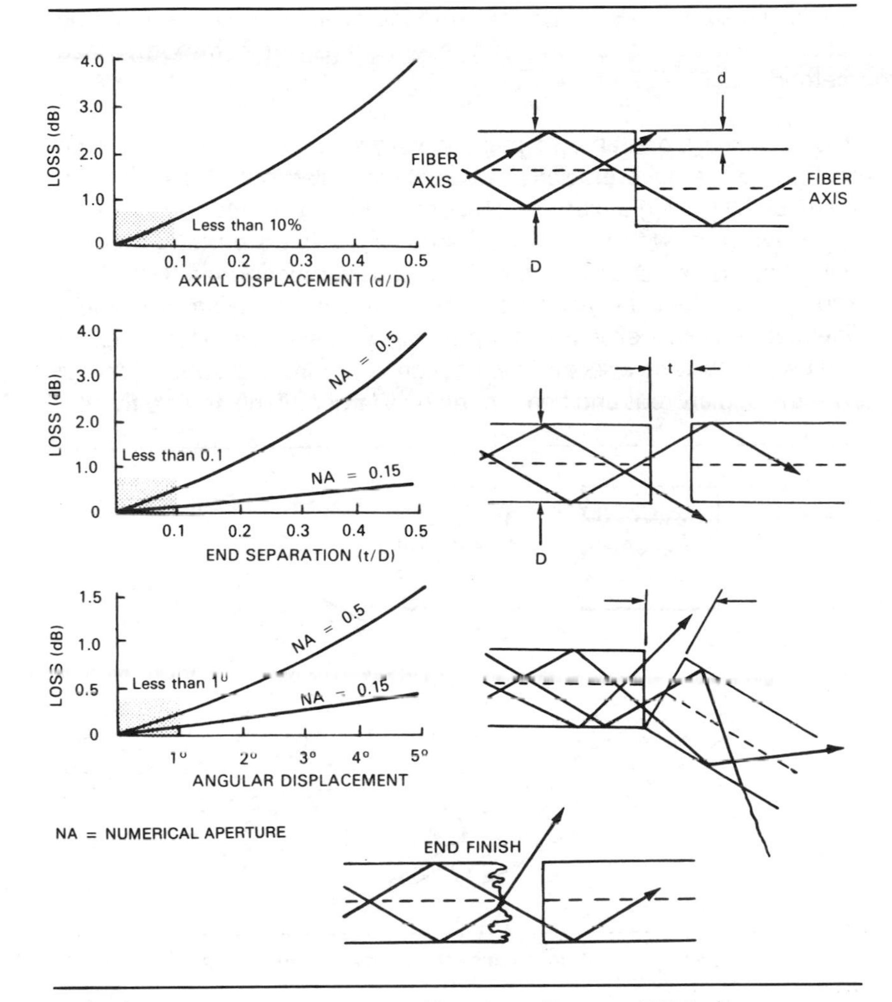

Fig. 3-10. Connector alignment considerations. (Courtesy AMP Inc.)

modulation can be applied to the light carrier wave by varying the amount of axial displacement, within limits, at the desired modulating frequency; by separating and bringing closer together the two fiber ends at the desired frequency; and by changing the angle of the splice from zero to the maximum allowable at the desired rate. Another possibility is in the use of a *Kerr cell*, a crystal that changes opacity (or, from a positive viewpoint, transparency) at the modulating frequency. Placed between two butted ends of the fibers, the cell would increase or decrease the amount of light being sent to the receiver. These modulating suggestions are not commonly used methods.

It is not always desirable to make a permanent splice, so the fusion method, while quite efficient, is sometimes ruled out. A coupler can be made by placing a piece of heat-shrink tubing around three or four alignment rods, with the polished ends of the fibers butted firmly together in the center of the triangle or square thus formed (Figs. 3-11 and 3-12). When the tubing is shrunk, a tight coupling is formed, adequate for most experimental purposes. An even stronger coupling can be had by using a second tier of alignment rods of greater diameter over the primary rods and heat-shrinking a larger tubing around these.

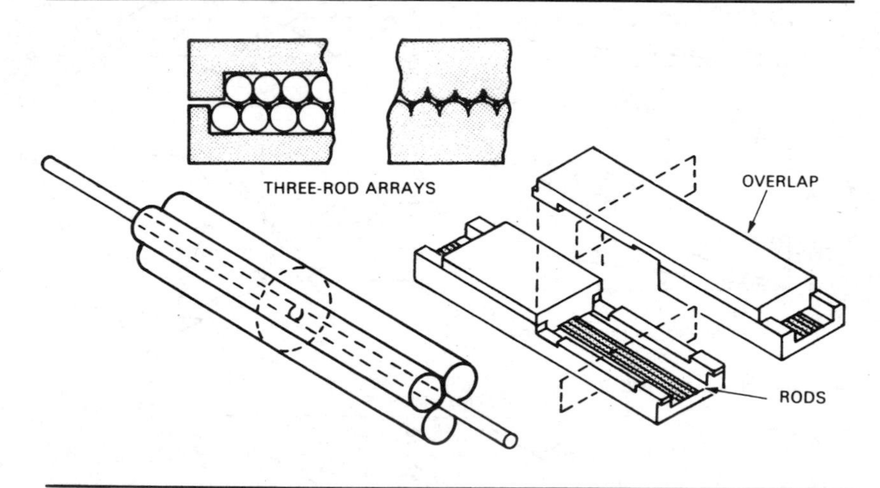

Fig. 3-11. Three-rod optical-fiber connectors, single- and multiple-fiber types. (Courtesy AMP Inc.)

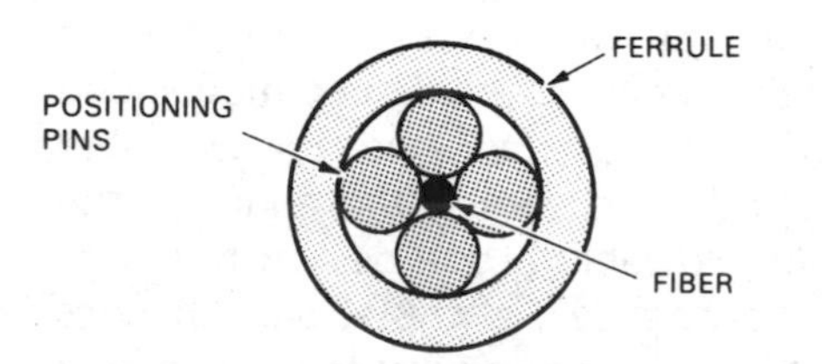

Fig. 3-12. Four-rod fiber-coupling method. (Courtesy AMP Inc.)

A V-notch can be cut lengthwise into each of two pieces of metal or fiber, which are then clamped face to face with the fiber firmly wedged in the Vs. For dual and greater numbers of fibers, V-notches can be cut for each fiber to be spliced. Blocks of these metal plates can be cut on both sides with V-notches, and the blocks can be piled one on top of another and firmly clamped, with a fiber in each notch.

Lathe-cut connectors employing concentric sleeves or cone alignment configurations can be made to order, but usually will be much cheaper to buy from a manufacturer. A partial listing of manufacturers is included in Appendix D.

Light Sources

Any light source within the frequency range acceptable by the fiber may be used. You can hold the end of a fiber up to a common household table lamp and view light exiting the other end many feet away (assuming that the fiber can propagate light in the visible range). If efficiency (total light power emerging from the fiber compared to the total light power at the source, Fig. 3-13) is a consideration, such a system would be out of the question. However, for experimental and basic purposes, efficiency can be neglected to achieve the main objective of the project.

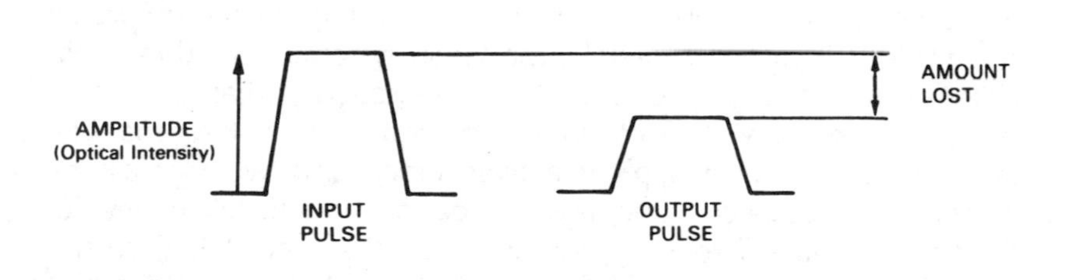

Fig. 3-13. Example of signal attenuation. (Courtesy AMP Inc.)

As experimental work becomes more sophisticated, it is necessary to take into account the wavelength of light best propagated by the particular fiber. As bandwidth requirements of the system become a limiting factor, the light source must be very carefully selected.

There are two major light sources in use in modern fiber-optics systems: the LED (Light-Emitting Diode) and the ILD (Injection Laser Diode). The ILD is considerably higher in price than the LED. The latter can be obtained from various sources for as little as ten cents each in odd lots, but will cost about two dollars each at retail, with specified performance ratings.

The more intense the transmitting light, the longer may be the length of fiber before a repeater is employed. More intensity usually means greater heat and greater physical size of the source. Placing light into the fiber at optimum angles requires the emitting area of the source to be smaller than the area of the fiber core. Further, the beam pattern of the source must be quite directional to accommodate the acceptance cone of the fiber (Fig. 3-14). The light must be as spectrally pure as possible to minimize the dispersion caused by different wavelengths traveling at different velocities in the fiber (material dispersion). The fiber material must likewise be pure and carefully manufactured to minimize losses from imperfections (Fig 3-15).

Digital systems require very short rise and fall times when the light source is alternately turned on and off; analog systems of modulation require the light source to vary linearly relative to the drive current. Taken together, these are very demanding requirements for an efficient, well modulated light source.

We might use a simple grain-of-wheat lamp to transmit intelligence over a fiber, but the delay it would have in turning on and off would severely limit its use in a digital information system. It could be modulated with a shutter, increasing response speed perhaps, but this, too, would be very slow compared to modern digital speeds. The same limitation would apply if a filamentary lamp were amplitude modulated by voice or music; it would be too slow to follow any but the slowest sound-frequency variations. On the other hand, such a source might be excellent for some purposes, such as simple signaling or simple feedback indicators. It would certainly be ample to turn on or off a remote power or light switch, or to indicate remotely whether a water tank were full or an air tank sufficiently pressurized.

Power Transmission

Sandia Laboratories in Albuquerque has experimented with transmission of power via fiber-optic cable. In some sensitive locations, it can be undesirable to carry power into and through an area in the usual

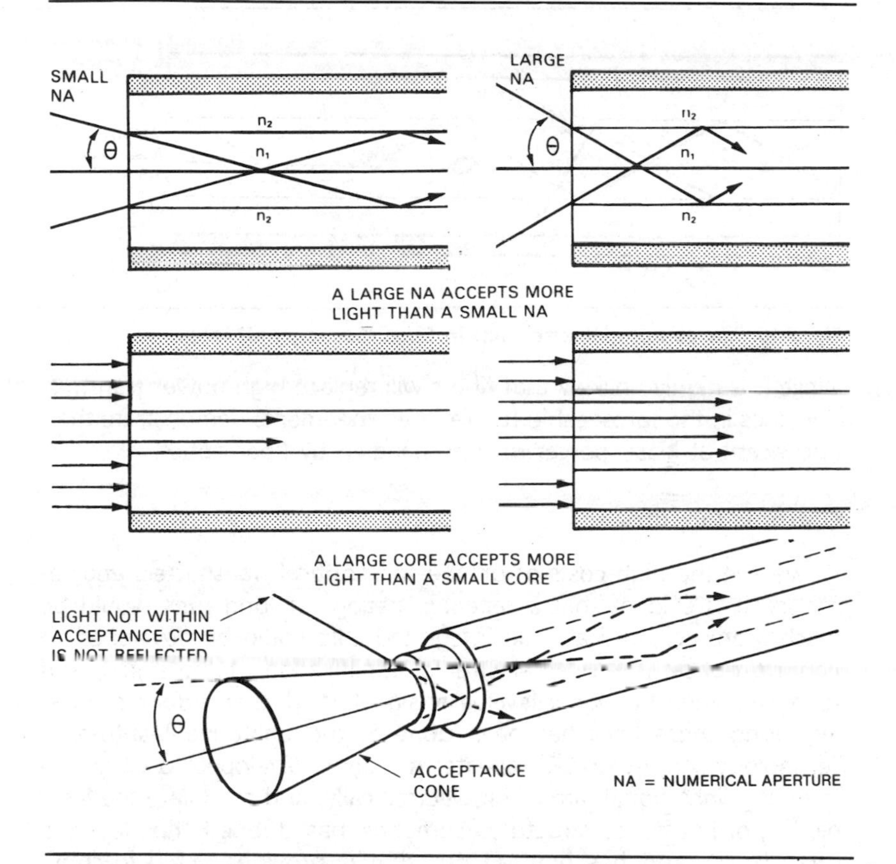

Fig. 3-14. Typical fiber acceptance cone. (Courtesy AMP Inc.)

copper-wire manner. Fiber optics may be a way of safely routing such power, although at present the efficiency of such a system is quite low.

Electrical power is converted to power at optical frequencies for transmission through fibers. At the remote location, the light is converted back to electrical power in a photovoltaic cell. Existing sources, detectors, and fibers handle only a few watts. Xenon-arc sources have an efficiency of about 5 percent. A gallium-arsenide photocell converts about 3.5 watts of optical power at an efficiency of about 23 percent. And yet, depending upon how critical a particular situation may be, even these low levels of efficiency may be justified.

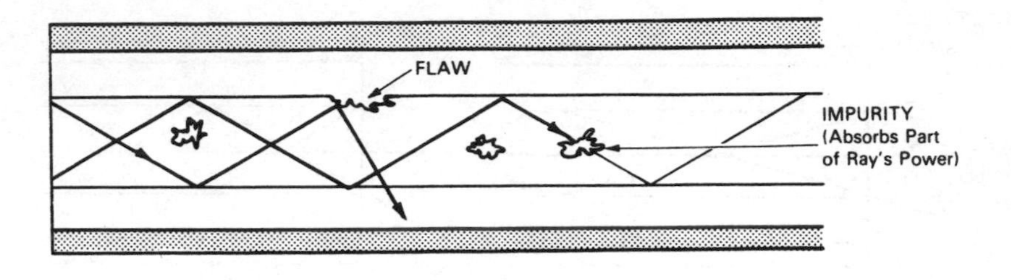

Fig. 3-15. Losses from imperfections in fiber. (Courtesy AMP Inc.)

While it is highly unlikely that fibers will replace high-power transmission lines in the foreseeable future, the experiments demonstrate that, in concept at least, power may be handled by fiber optics.

Optronic Circuitry

As we eye the high costs of off-the-shelf optical transmitters and receivers, it is evident that a repeater station for long runs would be much more economical if an integrated chip could be developed to perform optical receiving, electronic amplification of the signal, and retransmission of a higher-level laser signal. Work that is quite promising along these lines has been done at the California Institute of Technology. An integrated repeater has been developed to detect an incoming light signal, amplify it electronically, and modulate the laser diode portion of its structure. Someone has dubbed the device a "translaser" and has pointed out that it provides "chip-to-chip" communication.

In the future, other applications of optical/electronic (*optronic*) circuitry on a single chip may extend to include multiplexers and demultiplexers. Further into the future, we might envision integrated optronic logic circuits and signal processors.

4 The "Nuts and Bolts"

Light reflects to some degree from almost any surface. As any solar-panel enthusiast will tell you, it is difficult indeed to construct a surface that will absorb 100 percent of all the light that falls upon it. Some surfaces reflect some light wavelengths very well while absorbing others, and we say an object is blue, red, green, or some other color, depending on which wavelengths are reflected.

Some surfaces, even when highly polished, absorb certain wavelengths of light. Highly polished gold, for instance, still looks like gold because it does not reflect all colors 100 percent.

A mirror made of polished glass, coated with mercury or aluminum on its underside, will reflect all colors quite well, and for simple purposes such as shaving, it will serve nicely to provide feedback to the person reflected. For more exacting purposes, such a mirror leaves much to be desired because light is reflected from the front surface as well as from the "silvered" surface. Two images displaced in time and space by the thickness of the glass can make the rear-silvered mirror unacceptable for purposes such as range finders on expensive cameras. When a mirror is used in an enclosed, protected space, it is possible to coat and polish the front surface and thereby eliminate the double image.

A few of the millions of materials that reflect light from their surfaces will also convey light waves through their interior mass. Glasses are such materials, as are some plastics. Certain gems, such as diamond, are *transparent,* the common term for a material that conducts light quite readily.

A block of material can be *translucent.* It conducts light through its mass, but not as readily as a transparent material. Translucent materials usually scatter the rays and mix them in direction until the emerging rays are completely random, although they may have entered in a well-ordered "bundle." Translucent materials strongly attenuate light rays; that is, they reduce them in energy. The material absorbs part or all of the rays entering its mass, depending on the depth of the material and the intensity of the entering rays. Ocean water is an example. The absorbed energy appears in the form of heat, raising the temperature of the material.

A sample of material can be *opaque.* This is another way of saying that the light rays striking the surface of this material will be either almost totally reflected or absorbed within a few micrometers of the surface. Light will not propagate, or travel, within the mass of this object.

Basic Light Propagation in a Glass Bar

In fiber-optics work, we are interested in how well light will propagate through a given material (sometimes referred to as a *medium,* or the plural, *media*). We are also interested in how well or to what extent light will reflect from surfaces.

It is well known that light will reflect better from a polished surface if the rays approach the surface at an angle, rather than perpendicular to the surface. Solar panels on rooftops are sometimes fitted with clockwork or electronic orientation mechanisms to keep the panel exactly perpendicular to the sun both in azimuth (horizontally) and elevation (vertically). The very practical reason for this is that far more of the sun's energy will penetrate the glass surface of the panel if the rays strike perpendicular to the surface.

In ordinary glass and optics work, we are familiar with large plane surfaces, from huge store-front windows to tiny kitchen windows. In fiber-optics work, we will be dealing with glass and plastics, too, but now we will be interested in how well light travels end to end through the glass.

If a light ray strikes a flat polished surface at, say, a 45° angle instead of at the perpendicular, the ray will leave the surface at exactly the same angle in the opposite direction (Fig. 4-1). That is, the angle of incidence is equal to the angle of reflection. If the incoming ray is at an angle of 1°, it will leave the surface at 1° in the direction of travel, but altered 2° from its original direction. Some rays will travel along the surface of the material, but this is a special case for more intensive study in the physics of light.

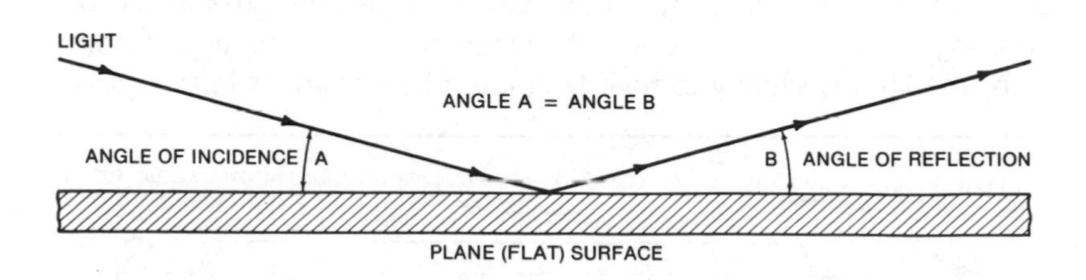

Fig. 4-1. Angles of incidence and reflection.

It is characteristic of light that when a ray is traveling in a medium such as glass, it will impinge on the boundary between that medium and another material such as air and will *refract* back into the original medium, to a greater extent the more shallow the angle of incidence. This is consistent with our premise that the greatest amount of light will enter a surface when the angle of incidence is perpendicular (90°) to the surface.

To refract, a ray of light must enter a medium and change direction, eventually exiting the material or refracting or reflecting from a second surface (Fig. 4-2). This action is in contrast to a reflective ray that simply "bounces" off the surface of the second contiguous medium.

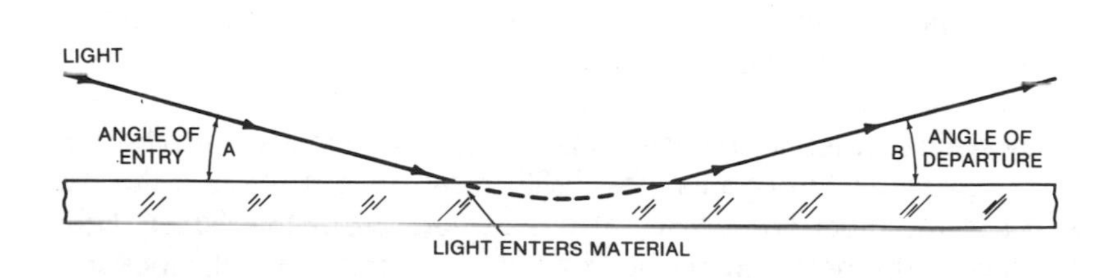

Fig. 4-2. Refraction of a light ray.

Suppose a long, narrow slice is cut from a pane of window glass, with the ends polished and exactly perpendicular to the long surface. We can examine the ability of the glass to carry light rays through its mass in the "long" direction (Fig. 4-3) rather than in the more familiar plane-to-plane direction. Place a source of light such as a tiny filament bulb at one end, and observe the light coming through at the other end of the strip of glass. You may be surprised to see the filament quite clearly, although there is obviously considerable extraneous light entering the glass from the four surfaces not of interest at this point. Also, the light is slightly to heavily colored by impurities in the glass.

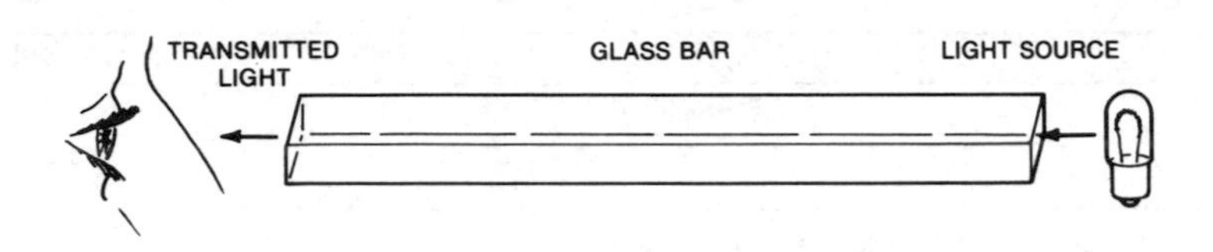

Fig. 4-3. Bar cut from heavy window glass.

There are ways of treating the four lengthwise surfaces to eliminate any interference from outside light. The two long, flat surfaces and equally long and flat edges may be painted black, or better still, polished and silvered to create internalized mirrors in four planes. Only the light entering an end would be seen at the other end.

A large, long square bar of glass would be quite expensive. Rather than go to this expense for experimentation, create an image of such a bar in your imagination. The ends are perfectly perpendicular to the long direction, and highly polished. The four long surfaces are also polished, and they are coated with aluminum molecules deposited in a vacuum, a treatment that is highly industrialized today. "Two-way" mirrors, among others, are made in this way, permitting store detectives to watch customers without being seen.

A ray of light entering the bar horizontally, exactly in line with the axis of the glass bar through its mass, will travel straight through the bar and exit the other end exactly at the center. This is the "ideal" light ray. Evidently there can be but one ray like this, with the possible exception of rays of the same wavelength but slightly out of phase. Uncountable other rays must take other paths through the bar.

It is impractical to make a light source that will emit but one ray in a given direction. A lamp filament, for instance, emits light rays in all directions, because the filament is quite large in comparison to the size of the ray of light. Even the light-emitting diode (LED), as small as its light source can be designed, is large in comparison to a single ray of light (Fig. 4-4).

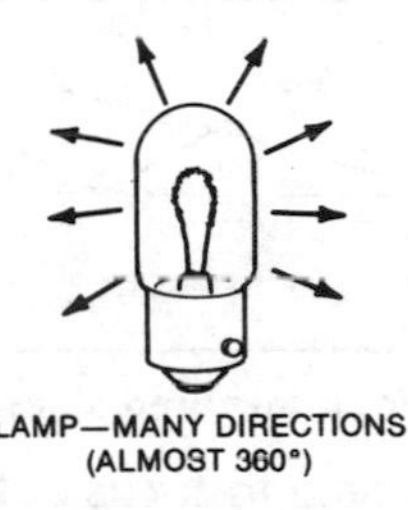

Fig. 4-4. Rays emitted from light sources.

For this reason, light will approach our long, square rod of glass at every possible angle, including the one lone ray that is axially perpendicular to the polished end of the bar. Let us look first at the off-angle ray that will enter at just the precise angle to propagate that ray to the inside center of one of the mirrored sides. There will be four of these rays, one to each center of the four inner surfaces (Fig. 4-5). The ray will be reflected at exactly the angle at which it approached, to exit at the point on the opposite end of the bar which will cause the ray to focus (come to a point simultaneously with other rays from other directions) at exactly the same distance from the exit plane that the light source is from the entrance end.

Another set of four rays will enter the bar at an angle which will reflect them from a point at exactly one-fourth of the source-to-focus distance, to a point on the opposing plane surface at exactly three-fourths of the source-to-focus distance, from which the ray is reflected again so that it focuses at the same point as the first ray described above.

There will be an infinite number of rays entering and an infinite number exiting. There will be, therefore, an infinite number of angles of incidence at which the entering rays strike the mirrored sides of the

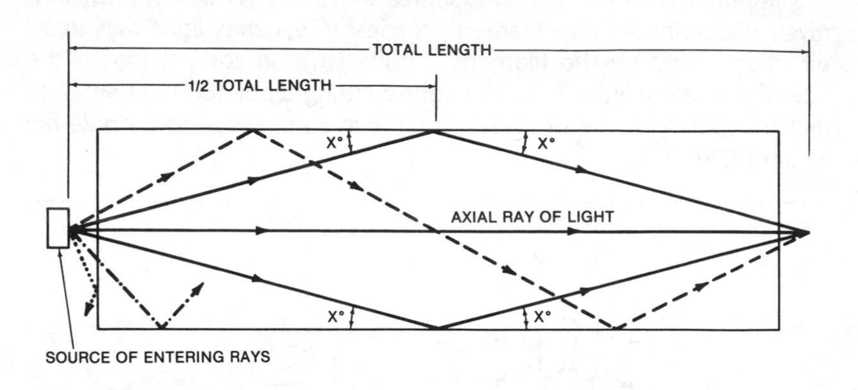

Fig. 4-5. Rays of light reflected internally from mirrored sides of glass bar.

glass bar. Many of these rays will not come to focus at the same point beyond the exit surface. Indeed, many of the rays leaving the light source will not enter the bar at all, but will be bounced off the polished surface of the source end of the bar.

Keep in mind that when light leaves one medium of propagation and tries to enter another of different *index of refraction* (a more precise way of stating that the two materials are of different composition), a certain percentage of the source rays will be reflected or refracted and will not enter. They are "lost," and this loss is one of the factors considered when computing the efficiency of a fiber-optics system.

Due to imperfections in the lattice structure of the glass crystal, and to impurities such as water and minerals in the glass, all of the rays are *attenuated* to some extent, as indicated in Fig. 4-6. (Such losses are sometimes called material dispersion.) Attenuation means, in effect, that the exiting rays are less intense than when they entered. The degree to which these rays are attenuated is a critical factor in the fiber-optics system, and fibers are comparatively rated to a standard, usually by a manufacturer's statement such as that used by Corning for their 1505 Optical Waveguide (glass fiber): "Attenuation @ 850 nanometers = 2.9 dB/km." A nanometer is one billionth of a meter. The "850 nanometers" is the optimum wavelength of light which this fiber will propagate. The shorthand symbol "dB" stands for *decibel,* a logarithmic measure of intensity of sound and electromagnetic energy. A kilometer is 1000 meters, and its abbreviation is "km." Thus,

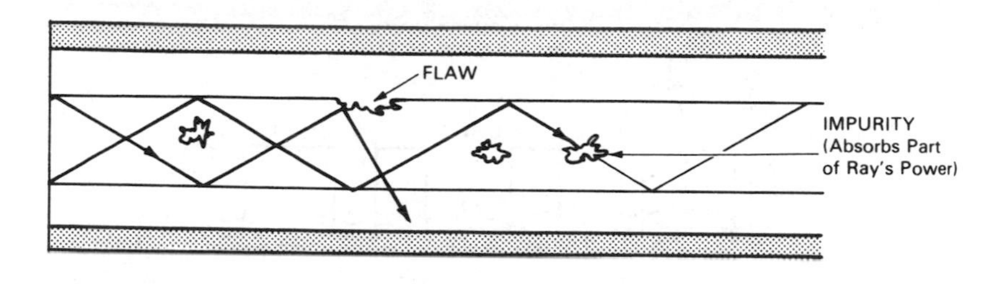

Fig. 4-6. Losses caused by fiber imperfections. (Courtesy AMP Inc.)

the attenuation of the Corning fiber is 2.9 decibels for each kilometer. Relatively speaking, this is an excellent figure of merit which indicates that, with a suitably intense source, the light will propagate perhaps as far as 30 or more kilometers (18 miles plus) before it becomes too faint to register above the noise level on a photoelectric sensor at the receiving end (Fig. 4-7).

Propagation in a Glass Rod

For practical reasons of manufacture, cabling, and propagation of light, square fibers are not used. Therefore, let us consider the bar of glass as being a cylinder with a highly reflective coating on its surface (Fig. 4-8). We now have a gigantic model of the filaments of glass employed in fiber optics. Instead of only four long, flat plane surfaces, we have an infinite number of slightly curved surfaces along the bar. Light rays leaving the source in all directions will approach the polished, transparent end, some to enter, some to miss the bar entirely, and others to be reflected from the surface.

Rays lost by reflection can be reduced to a minimum by treating the glass surface with an optical coating such as that applied to expensive camera lenses, but even with the best coating some loss will occur. If we draw a number of ray paths from the source to the bar end surface, we can readily see that a cone shape will be formed by those rays that will enter and propagate to the end. This is called the *cone of acceptance* of a rod or fiber (Fig. 4-9).

If the source can be placed closer to the surface, more light will enter the rod. The cone of acceptance can be improved somewhat by using

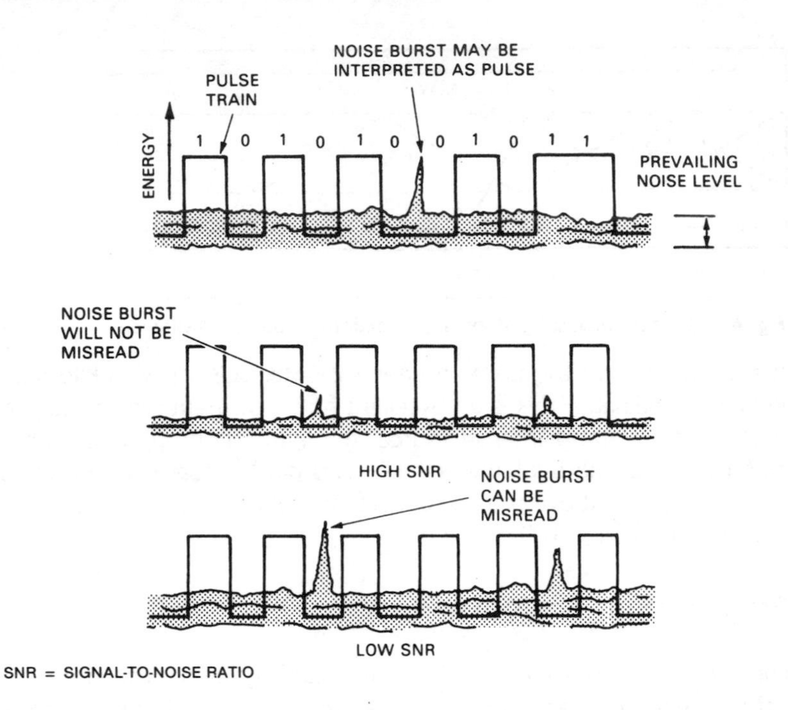

Fig. 4-7. The effects of noise in a fiber-optics system using a pulsed carrier wave. (Courtesy AMP Inc.)

a wetting agent or a system of lenses between the source and the end surface of the rod to capture more of the source light.

One of the best-remembered rules in high-school geometry is that "the shortest distance between two points is a straight line." The shortest distance from one end of our glass rod to the other is straight through the rod axis from one end surface to the other. Because the light source has only one point that corresponds precisely with the angle that propagates along the axis of the rod, all other rays from the source will enter at some angle less than the 90° angle of axis entry. This means, by applying the rule above, that all other paths will be longer than the axial path through the rod.

Light is electromagnetic in nature. Electromagnetic waves travel at a finite speed in a given medium, but not at the same speed in all

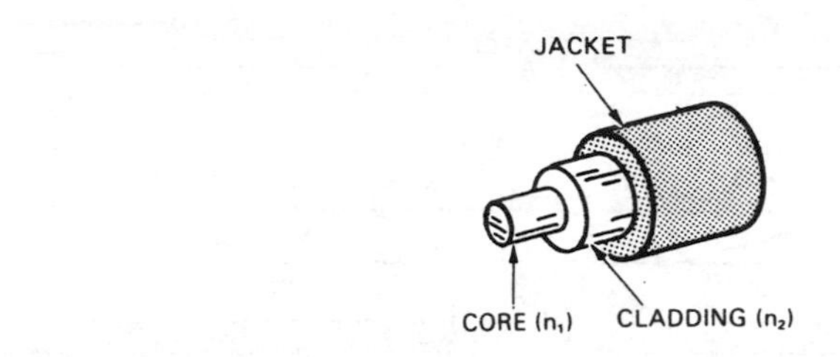

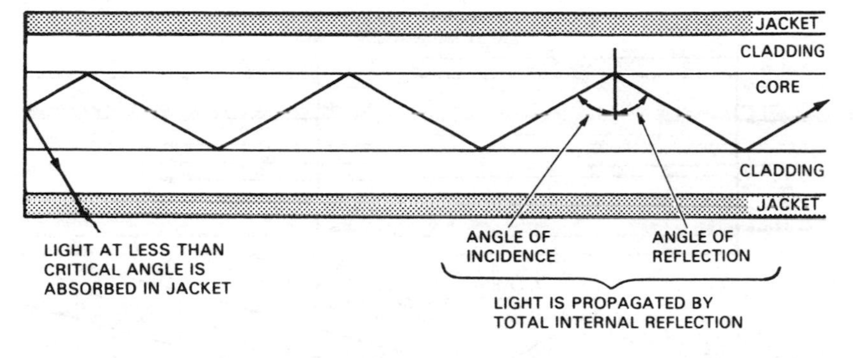

Fig. 4-8. Cylinder of glass as model for fiber. (Courtesy AMP Inc.)

media. If the path for a given ray is longer than the axial path, that ray will take longer to travel through the rod than an axial ray. Differences in travel time adversely affect the system at the receiving end, causing some distortion of the *signal.* (The signal is the light ray with information superimposed upon it.)

By reducing the diameter of the glass rod to the fineness of a human hair, a worthwhile reduction in the distortion caused by differences in travel time can be achieved. There are, however, other sources of degradation to be considered.

In camera and telescope lens-making, there is a source of distortion known as *color aberration.* If not corrected, it causes blurring of an otherwise sharp photograph. In the world of electronics, color translates to differences in the frequency of light. By using a laser source (which is single-color, *monochromatic,* light), a single-frequency ray of light may be transmitted through the fiber. However, when the source is *modulated,* that is, when intelligence is superimposed upon the ray of light by altering its frequency or intensity at the information rate,

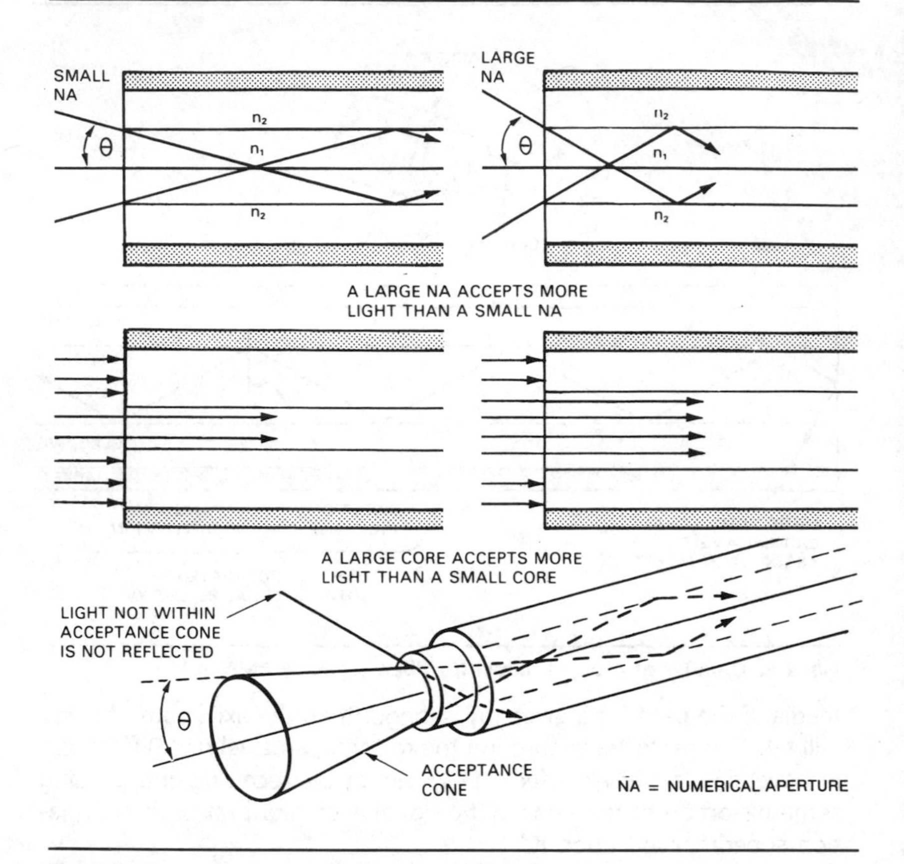

Fig. 4-9. Cone of acceptance. (Courtesy AMP Inc.)

this creates *sidebands,* or added frequencies and subtracted frequencies. In effect, this changes the color of the light beam, in the case of light in the visible spectrum. (It can be argued that you do not change the color of invisible light; however that may be, the frequency of the ray is changed.)

Safety Precautions

Light propagation in a fiber is not limited to visible light. Depending on the composition of the fiber and its degree of purity, electromagnetic waves both above and below the narrow band of frequencies that affect the human eye can be propagated. Therefore, a hard-and-fast rule should be memorized and followed strictly: NEVER LOOK INTO THE END OF AN ENERGIZED FIBER. A most advisable and equally beneficial corollary is: NEVER LOOK INTO THE WINDOW OF AN ENERGIZED LASER DIODE. You would not defeat the interlocks on a microwave oven and hold your hand in its cooking space while the timer ticks away—to do so would be to cook your bone marrow and flesh before you could pull your hand out of danger. Microwave energy is no less active because of its invisibility. Likewise, invisible light energy can destroy your vision just as surely as though you stared into the sun without protection.

5 Basic Principles of Glass Fibers

Light traverses a vacuum at a speed of 299,806 km per second (186,291 miles per second). Its speed in any other medium is slower. The ratio of its speed in a vacuum to its speed in another medium is the *index of refraction* of that medium. In optical glass, the speed of light is 198,548 km/sec (123,372 mi/s). Divide the speed of light in glass into the speed of light in a vacuum, and you see that the index of refraction of glass is the ratio 1.51:1, or simply 1.51.

Air and water are two quite common adjacent media that are sometimes fairly good carriers of light. We have learned through the years that when we see an object in the water, it appears closer to the surface than it is when we reach for it (Fig. 5-1). Water has a different index of refraction than air, and therefore light rays will be bent at the surface between the two media.

The index of refraction is not constant for all light wavelengths. Blue (shorter wavelength) is refracted more than red (longer wavelength), for instance. This accounts for the effect of a prism (shown earlier in Fig. 3-6) when white light is passed through it. Flat, parallel-surface glass will also separate light into specific colors, but because of the reverse action of the second surface, the beam is restored to white light.

Referring once again to our long glass rod, if we silver the cylindrical surface of the rod, light striking that surface from inside will be reflected back into the rod in the forward direction of the traveling ray (or wave, as it is also called). As much as 98 percent or more of the

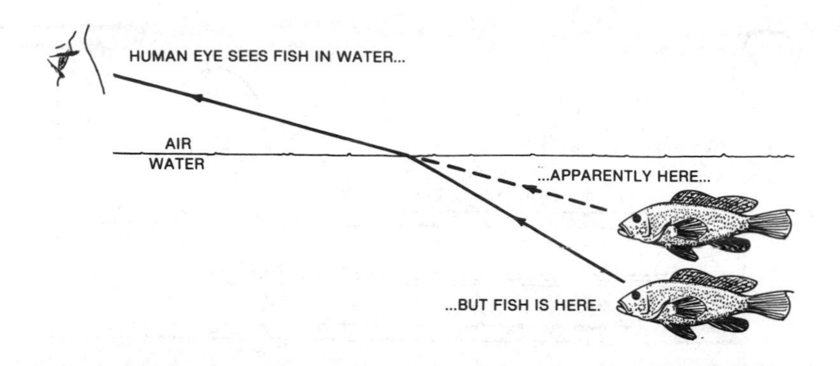

Fig. 5-1. Air-water interface.

light will be reflected, depending on the quality of the mirrored surface and the sharpness of the angle at which the light approaches the surface. There are other ways of causing the light ray to be reflected or refracted back into the desired path. The rod can be coated with a layer of plastic of different index of refraction than the glass, and the layer will refract an incident ray back into the glass rod. Depending on the material used for the coating and the sharpness of the angle of incidence, the cladding (coating) can be as effective as though highly polished silver were used.

If the thick rod is heated and drawn into a long, thin fiber, light will travel from end to end just as before—much better, in fact. If we clad the fiber with a glass or plastic covering, virtually all of the light entering one end will travel down the fiber and emerge at the far end, because it is either reflected or refracted from the surrounding medium. This is the beginning of fiber optics.

Since this is a book primarily of experiments and projects, we will omit many of the more intricate details of how fibers transmit light and get on with the experiments. First, however, it is well to understand the difference between *step index* and *graded index* fibers. Step-index fibers are similar to the rod described above, but with cladding of more than one material (Fig. 5-2). There is a sharp line of demarcation between the core medium and the cladding media. Graded-index fibers have a unique method of manufacture that causes the refractive index to decrease continuously with radial distance from the fiber axis.

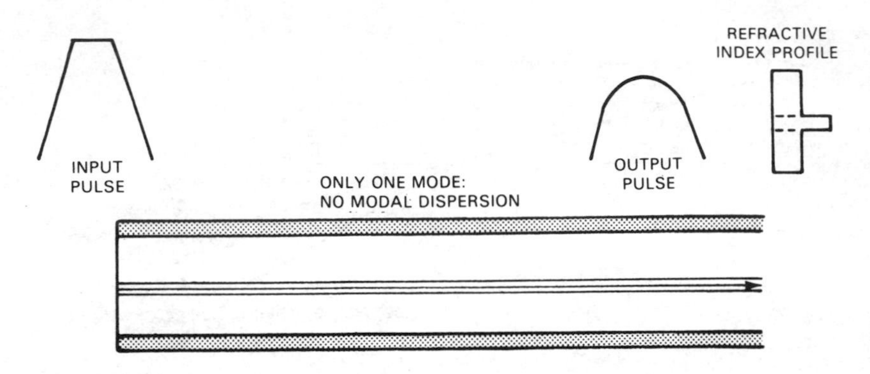

ADVANTAGES	DISADVANTAGES
Minimum Dispersion	Small Numerical Aperture: Requires Laser Light Source
Large Bandwidth: High Operating Speeds	Difficult to Terminate
Very Efficient	Expensive

Fig. 5-2. Single-mode step-index fiber. (Courtesy AMP Inc.)

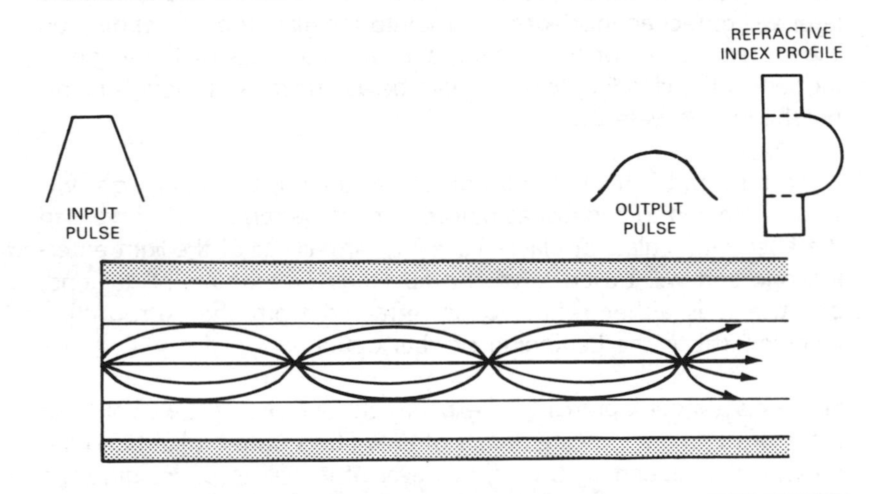

ADVANTAGES AND DISADVANTAGES
The Multimode Graded-Index Fiber Is Intermediate between the Other Two Types; Its Advantages and Disadvantages Lie between the Other Two.

THE FIBER IS SUITED FOR MEDIUM DISTANCES AND MEDIUM OPERATING SPEEDS.

Fig. 5-3. Multimode graded-index fiber. (Courtesy AMP Inc.)

The effect of graded-index fibers on a light ray trying to leave the axis is to increase the speed of the ray the more distant it is from the axis. At the same time, the ray is refracted back toward the fiber axis. The end result is that a ray traveling a longer route will arrive at the far end at the same time as the axial ray (Fig. 5-3). Step-index fiber is illustrated in Fig. 5-4.

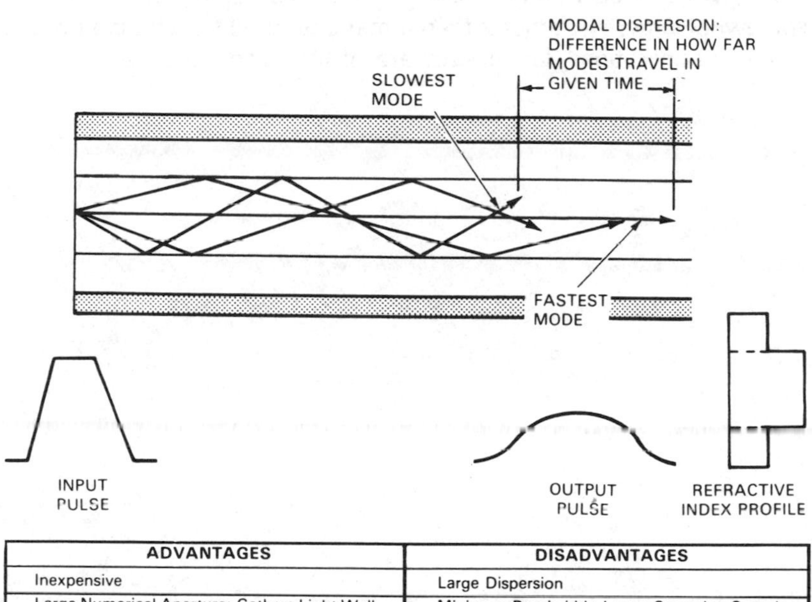

ADVANTAGES	DISADVANTAGES
Inexpensive	Large Dispersion
Large Numerical Aperture: Gathers Light Well	Minimum Bandwidth: Lower Operating Speeds
Easy Termination	

Fig. 5-4. Multimode step-index fiber. (Courtesy AMP Inc.)

There are many types of fibers available. Most are sold on reels of upward to a kilometer in length, and longer. Manufacturers have minimum-shipment order requirements, so it is impractical to order direct from Corning, for example, unless you have an application in mind for their cable that will require several kilometers of fiber. Even then, you may be referred to one of their agents, or "manufacturer's representatives," as they are commonly called.

Shorter bundles and lengths may be obtained from various surplus sales houses. A few retail mail-order houses have new fiber available in convenient bundles and lengths, although exact specification requirements may be subject to considerable searching on the experimenter's part. A few known sources are listed in Appendix D.

Depending on the use to which the fiber will be applied, there may be some interesting alternatives to buying manufactured products. One of the first experiments to follow uses a flexible tubing filled with water, which works fairly well in its intended setting. It is obviously somewhat larger in diameter than a manufactured fiber would be, but in many applications, larger sizes are of little consequence.

6 Simple Systems

Now that you have the peculiarities of light-carrying fibers well in mind, it is appropriate to consider how information can be conveyed from one point to another via light. The basic fiber-optic method is shown in block form in Fig. 6-1.

Basic Methods of Signaling With Light

Light has been used as a medium of communications for centuries, perhaps millenniums. A lantern on the mast of a ship traveling at night warns other crews and their captains that they have company approaching over the horizon. A red light on the port (left) side of the ship and a green light on the starboard (right) side, with a white light on the mast high above, communicate to each navigator the direction the other ship is traveling.

Aircraft are required by law and the application of common sense to carry port and starboard lights when flying at night. The white light is usually rotating, and sometimes it is a brilliant flashing light called a "strobe." Pilots can tell instantly whether they might intersect the course of another plane in the same area simply by "reading" the red, white, and green lights.

Ships can communicate at night in the old-fashioned manner, by opening and closing a shutter on a searchlight in the International Morse code pattern. There are standard abbreviations in an international code book carried by all ships. These abbreviations promote surprisingly rapid communications in all languages by "blinking

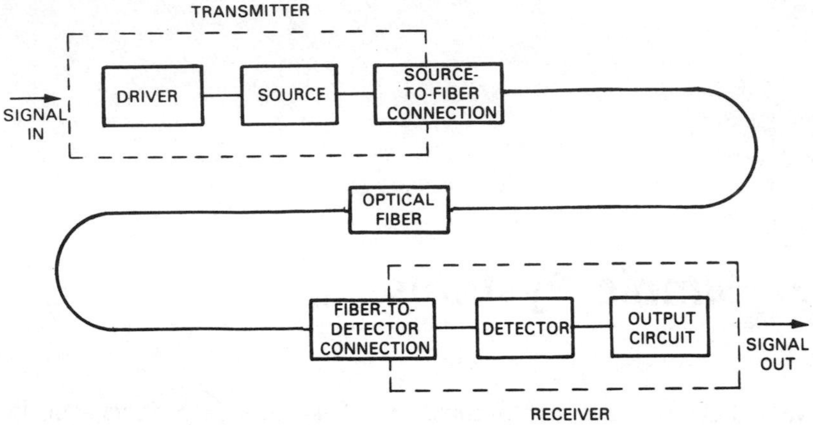

TRANSMITTER
SIGNAL IN
DRIVER
SOURCE
SOURCE-TO-FIBER CONNECTION
OPTICAL FIBER
FIBER-TO-DETECTOR CONNECTION
DETECTOR
OUTPUT CIRCUIT
SIGNAL OUT
RECEIVER

Fig. 6-1. Block diagram of basic fiber-optic link. (Courtesy AMP Inc.)

lights." It is not uncommon to see the yardarm lights of Navy ships in the harbor blinking in code at night; the powerful beams of searchlights are not needed at such short distances. Blinking lights are used to send messages locally to keep the radio spectrum as clear as possible when ships are in port, leaving the radio communications medium for ships at sea.

The Navy also uses other, less obvious means of light communication that can convey voice commands and general tactical information from ship to ship over line-of-sight distances. A long tube with an infrared diode (LED) at the focal point of a concave mirror is pointed toward a similar tube at the receiving ship. The LED is voice-modulated. The invisible beam travels through the air and excites a photodiode on the receiving ship. By this means, the ships' captains can converse with one another at sea or send tactical traffic, without risk of interception by an enemy.

An obvious question immediately arises: how can the source and receiver be kept in alignment while ships are rolling and dipping? This is done by diverting a tiny portion of the transmitted beam to a trio of orientation photodiodes; the amplified outputs of the photodiodes are applied to azimuth and elevation motors. Of course, the system is not used during thick fog or other adverse conditions.

We have mentioned three means of conveying information by light: steady lights (including colors), intermittent lights turned on and off in a recognizable pattern, and voice-modulated lights.

The Morse-code method of modulation is the forerunner of what is now called *digital* modulation. For code transmission, the light is either off or it is on. In Morse code, the length of time the light is on is either "short" or "long." Three short bursts represent the letter "S." Three long bursts represent the letter "O." Three shorts, three longs, and three shorts represent "S O S," the international distress signal. Variations of these letter codes and certain abbreviations such as "Q" signals (groups of three letters beginning with Q) make possible simple communications by means of light. For example, "QRM" stands for "I am being interfered with on this frequency" in the English language. The same meaning can be expressed in any language, and with a standard table of Q signals, navigators, captains, and amateur radio operators can communicate with anyone in any language, as long as they stick to the meanings for the Q signals in their own language.

Communication With Pulses

If the light beam is turned on and off at a fast enough rate, human vision loses its ability to differentiate and sees only a blurred on condition of the transmitting light. This fact is used to advantage in showing movies and television; the latter is produced on US screens at 30 frames per second, resulting in what looks like a solid picture. Movies are shown by blanking the screen during movement of the film from one frame to the next, resulting in 24 frames per second of light on the screen.

Light beams can be turned on and off at rates in the millions of "blinks" per second to produce digital pulse signals. There are at least nine different methods of modulating digital pulse signals; Fig. 6-2 shows three of the basic methods. In *pulse-rate modulation (prm),* the rate at which pulses of equal amplitude are generated varies with changes in the modulating signal. In this example, the rate increases when the amplitude of the modulating signal increases; the rate could decrease with increasing amplitude just as easily. In *pulse-width modulation (pwm),* the width (duration) of the pulses increases (or

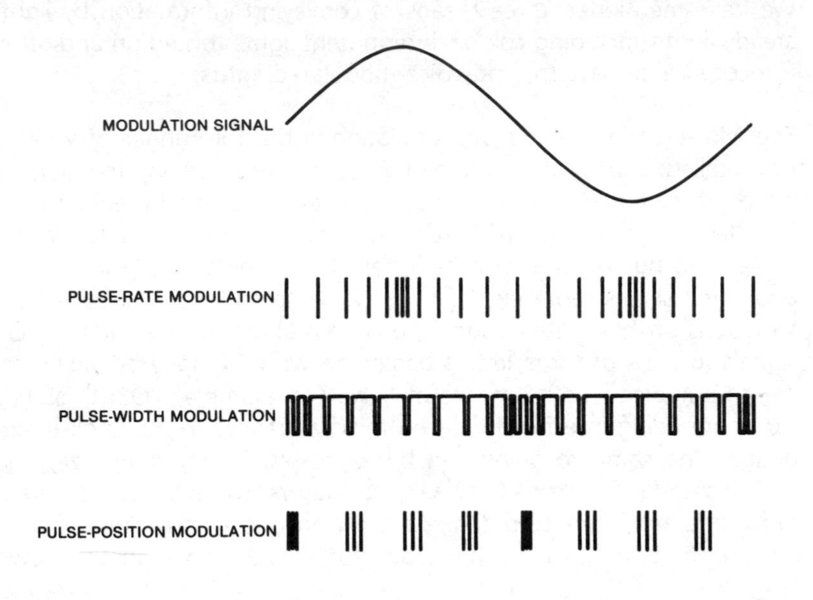

Fig. 6-2. Three methods of modulating digital pulse signals.

decreases) with an increase in the amplitude of the modulating voltage. In *pulse-position modulation (ppm),* the position of a pulse or series of pulses with respect to a reference clock pulse is varied in accordance with the modulating signal.

Another method is *pulse code modulation (pcm).* The fundamental idea of pcm is depicted in Fig. 6-3.

Multiplexing

Because of the need to use expensive lines efficiently, telephone companies have made *multiplexing* a common method of handling digital signals. Multiplexing is the apparently simultaneous transmission of multiple channels over a line. However, when they are examined minutely the transmissions are not precisely simultaneous. Tiny bits of voice samplings are sent alternately through the line by making use of the "quiet" instants that exist in all conversations.

Consider a movie film being projected on the screen: there is about as much "darkness" as there is picture, because the light is cut off by a

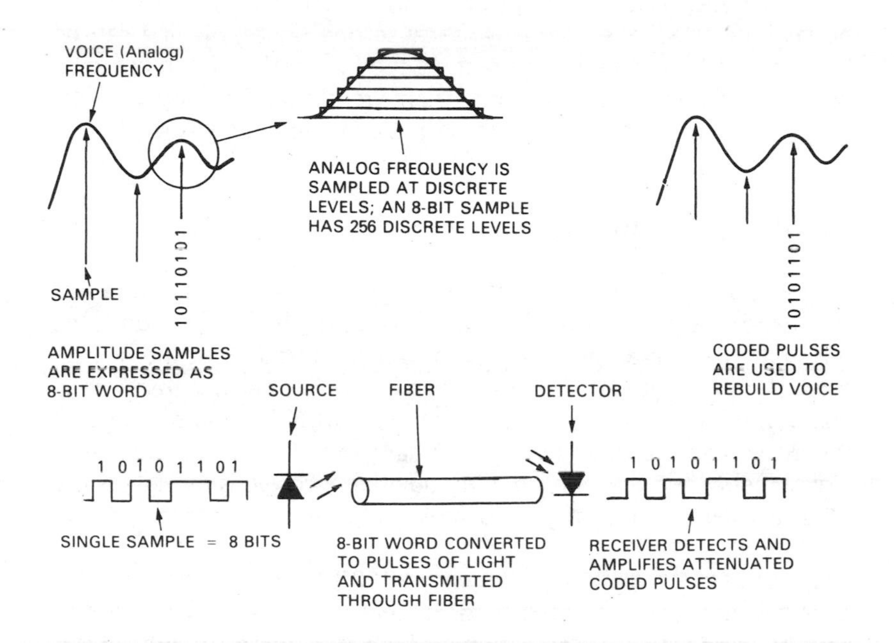

Fig. 6-3. Pulse-code modulation in relation to the amplitude of the modulating envelope. (Courtesy AMP Inc.)

shutter each time a frame is moved into place. If two synchronized projectors are used (or a single projector with double-width film), it is possible to project "stereo" movies, using the blank period for one projector to project the picture pulse for the other side of the stereo view. One projector is fitted with vertically polarized filters and the other with horizontally polarized filters; the observer uses polarized glasses to view the movie. (In another version of stereo movies, one projector uses a nonpolarized green filter and the other a non-polarized red filter; the observer uses glasses with one red lens and one green lens.)

In pcm, samplings from a person's voice in Detroit, for instance, are sent through the fiber to Flint, there being at least as much "unused time" between pulses as there is time occupied by pulses. The same fiber can be used for transmitting in the other direction, from Flint to Detroit, by a person unknown and unheard by the first two parties, with both conversations going on *at the same time.* The "unused time'' between all parties is put to work, and everyone is happy, especially the telephone company, because one fiber-optic line can do the work of two or more. In practice, the ''unused time'' in any conversation is considerably greater than we realize, making possible multiplexing of more than two transmissions simultaneously.

Digitizing Frequency

In modulating a light beam digitally, the pulses must occur at no less than twice the rate of the highest frequency of the information to be transmitted. For example, if we wish to make certain that the highest musical notes that can be sensed by any human ear are transmitted digitally, we must pulse the carrier at least 40,000 times per second. When we consider that television is sent over fiber-optics lines in a digital mode, and that the video modulation involved is as high as 4.5 MHz, it is evident that digitizing must occur at no less than 9 MHz.

If we increase the digitizing frequency to 600 MHz, we can divide 600 by 9 and find that we could send 66 channels of television over this light beam. Because of the extremely high frequencies of light, this is not an impractical situation in fiber optics. Thus, we begin to see the almost incredible potential that exists in using communication with a beam of light.

Simple Amplitude Modulation

The beginning experiments to follow involve simple signaling and then amplitude modulation of a beam of light. The basic forms are shown in Fig. 6-4.

Suppose we connect a carbon button microphone from an old telephone in series with a small flashlight bulb, and apply a dry cell of sufficient size to cause the lamp to light to about half its usual brightness with no sound input (Fig. 6-5). We have made a more or less practical, though simple, amplitude-modulation system. As we speak into the microphone, the carbon button causes greater or less resistance to be inserted in series with the battery and lamp, at the same rate as the changes in our voice amplitude. Within certain limits, the lamp becomes bright and dim, following the changes in our voice.

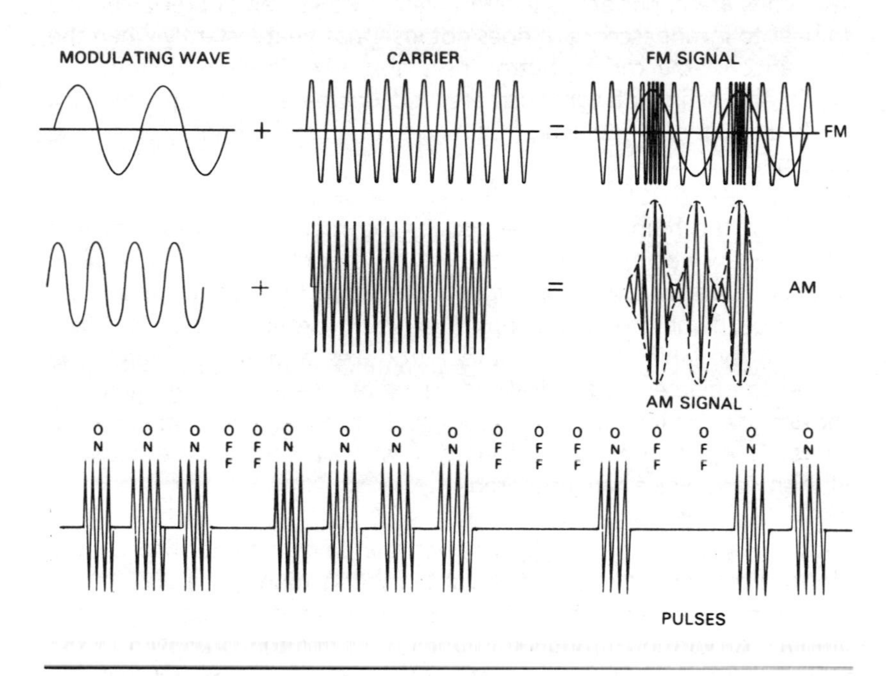

Fig. 6-4. Carrier wave with amplitude (am), frequency (fm), and pulse modulation. (Courtesy AMP Inc.)

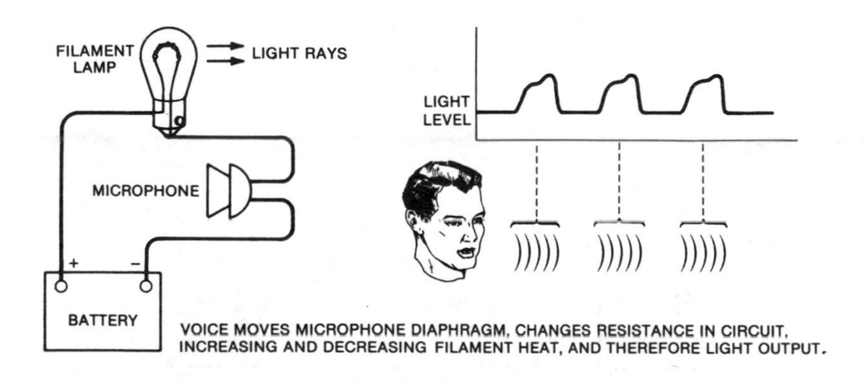

Fig. 6-5. Simple filament-lamp analog modulation system.

The limits are important. The lamp filament takes an appreciable time to heat to incandescence. It does not lose that heat instantly when the resistance of the carbon button rises abruptly. There is, therefore, a lag in the action of light increase and decrease in a filament. This limits the higher modulating frequencies that can be impressed on the light beam.

However, the human voice, especially the male voice, has sufficient low-frequency components to allow the system to operate intelligibly. Smaller, lower-wattage bulbs follow the voice to a higher pitch than do larger, higher-wattage lamps. We would expect to use the finest filament available, then. But this introduces another problem: The lighter the filament is physically, the less intense is the beam, and the shorter the distance that a medium such as air or fiber can carry the beam. Our choice of filament lamp must be a compromise between intensity and voice range desired.

Fortunately for the field of fiber optics, the development of solid-state electronics brought with it the light-emitting diode, the LED. In essence, this is a source that generates light in linear proportion to the amount of current through the junction. If voltage is applied in the forward direction to the LED, the intensity of the emitted light increases in direct proportion to the increase in voltage. All that is necessary, therefore, to modulate a light beam with a LED is to vary the voltage applied to the LED at the modulation rate. This is quite easily accomplished. The carbon button from a telephone handset will modulate almost the full range of the human voice if connected in series with a battery and LED (Fig. 6-6).

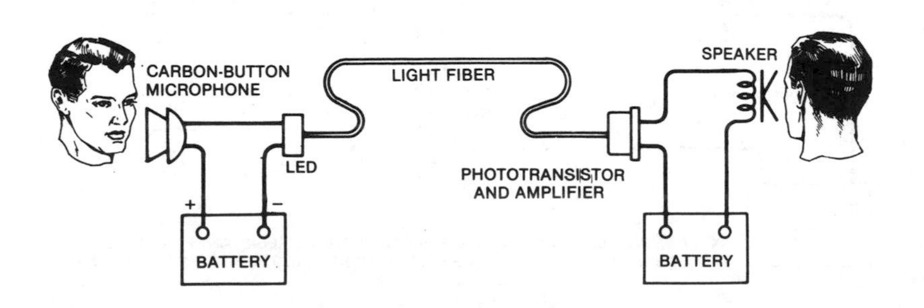

Fig. 6-6. Simple fiber-optics system that will work.

Carbon microphone buttons are not the best for this purpose for a number of reasons. Instead, let us use a dynamic microphone, either purchased for the purpose or made from a small speaker. Fairly good microphones can be made from one unit of a good stereo headphone set, or from a small dynamic speaker. The microphone output is boosted by a transistor amplifier, and the amplifier is matched to the LED. Light from the LED may be transmitted through the air, through a plastic tube filled with water, or through a glass fiber.

Sensing Devices

At the receiving end, there is need for a means to sense the received light. Fortunately, a number of photosensitive devices are available.

Photosensitive vacuum tubes have been in use for decades, as sensors for sound-on-film movie projectors and many other industrial and experimental uses. These require a steady voltage source of at least 90 volts. Photomultiplier tubes are also well known. These have a number of photocells in cascade within an evacuated envelope, and they require a number of stepped voltages, one to each unit, in increasing increments.

There are devices such as the plates of old selenium rectifiers that will work, but quite poorly for communicating purposes. (These can be used for solar cells to change sunlight to electricity.)

Cadmium-sulfide cells are photosensitive, but like the filament lamp they have a relatively slow response to changes in light level, and their usefulness is therefore limited to simple signaling devices. There are many of them around, and they are quite inexpensive, so if the application warrants, there is every reason to use them.

Solid-state physics has given us some highly sensitive photocell units that are excellent for use in fiber-optics work. There are texts treating their theory of operation, which we will not go into here. The price of these units has dropped considerably over the years, until they can be obtained from many retail and mail-order houses for as little as 19 cents. However, in the interests of reliability and knowing exactly what you are using, manufacturer-guaranteed units may be employed. They can be obtained from such sources as Sylvania for about two dollars.

Solid-state units equivalent to the vacuum-tube photomultiplier may be had as well. These are called *phototransistors* and *Darlington phototransistors* and are somewhat more expensive. However, they are obtainable from the sources mentioned above.

Both source and receiving units should be selected carefully to fit the needs of the system being built. Corning's optical waveguide 1505 operates best in the nominal 850-nm band and with a bandwidth at this midfrequency of 39 MHz. At 140 μm it is slightly larger than a human hair and will carry roughly eight television channels simultaneously.

Cable TV Applications

Applying more than one television channel to a light beam is made possible by multiplexing the signal at the transmitting end and demultiplexing at the receiving end. It is not nearly as difficult to accomplish as it may sound.

Experimenters interested in the practical "how-to" in this regard would find a tour through the rack room of a modern cable-television company quite instructive. While most cable-tv systems still use concentric cable, many new installations are being made with fiber-optics cables constructed as in Fig. 6-7. The principle of modulation is quite similar in either case. From 12 to 24 or more channels are multiplexed on the single cable and sent out to all subscribers. At the receiving end, the cable-television customer's "down converter" demultiplexes the signal so that the customer can use his regular television receiver on the cable, changing channels as though using an antenna and receiving the signal "off the air."

It should be mentioned that tuning channel 4 on the cable will not necessarily bring the same channel on the screen as appears at the open-air antenna, direct from the local television station. The cable-system multiplexer can put any particular incoming channel on any cable channel desired, and this is often done.

In addition to the regular channels, other systems, sometimes called pay television, are coming into use. These use *scramblers* to distort the modulation or the synchronizing pulse of each frame in a predeter-

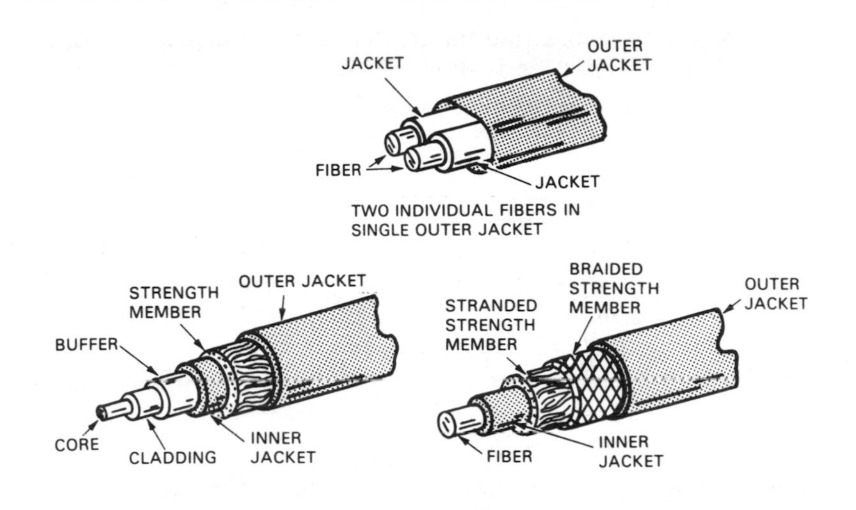

Fig. 6-7. Variations in cable construction. (Courtesy AMP Inc.)

mined pattern, requiring a paid-for service to "unscramble" the signal before it is applied to the customer's television receiver. (While the legality of doing so is under attack, many technicians know how to descramble such signals and can pick them off the cable by building their own "black box.")

All these services will probably "go fiber" sooner or later. The advantages have been mentioned already, and the disadvantages are so few that designers of new cable installations may be persuaded to use fiber without waiting for the systems to decrease in cost.

Booster Amplifiers

No matter which system of modulation is used on the basic light beam, after the beam travels a given distance through the fiber, its intensity will become attenuated to the extent that it will not reliably operate a photosensitive receiver. For this reason, before the distance becomes too great, a *booster amplifier* is inserted in the transmit-receive path.

Because of the possibility of increasing distortion of amplitude-modulated light with each reamplification, boosters are not ordinarily used in this mode. When booster amplifiers are needed, it is better to use digital modulation at the transmitter, because digital signals can be picked up, reshaped, and retransmitted in original form and intensity with virtually no introduction of distortion to the original signal.

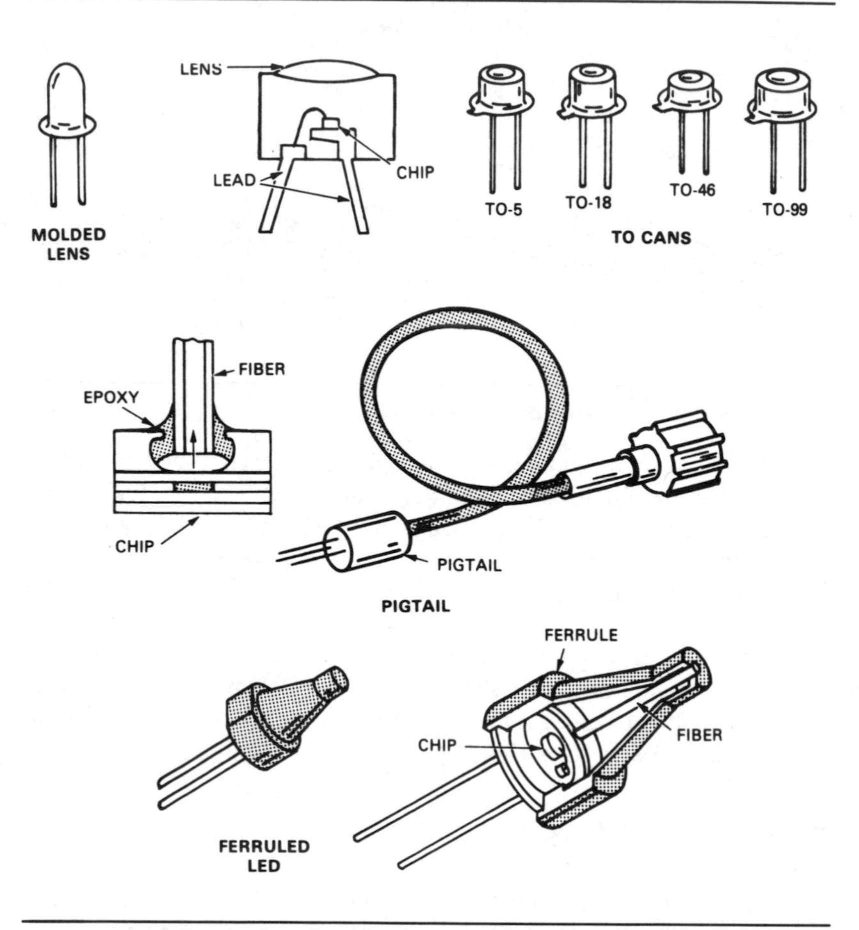

Fig. 6-8. Various forms in which photosensitive devices and LEDs are available. (Courtesy AMP Inc.)

Cost and power source are the only major considerations in deciding how many boosters are practical for a long-line fiber-optics system. When the original fiber-optics cable is manufactured with a minimum attenuation rate, the number of boosters required in the system is reduced accordingly.

Another source of attenuation in the case of booster insertion is the connectors that are required. Coupling always causes a greater loss than occurs in an unbroken fiber. However, manufactured couplers are becoming more and more efficient, and in the very long-line installations it is possible to weld the fibers together and insert them directly into the photosensitive diodes with a tiny drop of matching fluid. In fact, the devices can be manufactured into couplings, as shown in Fig. 6-8. In this manner, losses can be kept lower than one-half decibel per booster. The loss caused by couplings does not degrade the signal; rather the number of boosters across a given long-line installation is increased. Our experiments cover a few of the many coupling methods, some of which are used only experimentally.

Section 2
Experiments

7 Experiment 1: Getting Acquainted With a "Light Pipe"

This and following experiments contribute to hands-on knowledge of how the principles of fiber optics can be applied to ordinary materials. Quite possibly, once you become familiar with the practical aspects of working with light communications, you may come up with one or more practical uses for the completed experimental units.

Discussion

It is not axiomatic that because you can see through water and other liquids these are good conductors of light. A window pane is only an eighth of an inch thick, but the same glass from which it is manufactured might be worthless for making optical fibers. Water that looks crystal clear in a brook might well be so contaminated with air, minerals, and bacteria that light will penetrate only a few feet. In your experiments, you can get a feel for what can be done to pass light through water by using distilled water purchased from the supermarket. Your "light pipe" will carry light a few hundred centimeters with fair success; assess its ability to carry red, green, yellow, and infrared light, and see which performs best.

The cadmium sulfide (CdS) cell may be purchased from companies advertising in electronics and hobby magazines, or it may be salvaged from an old television receiver that employs automatic brightness control. Old photographic light meters, now thrown out with yesterday's newspapers because modern automatic cameras have made them obsolete, sometimes contain these cells.

Read through the materials list and procedure steps before performing any actual work. This will give you an overall view of what the experiment requires and enable you to take advantage of certain steps, such as when drying cement, to perform other work necessary to continuing.

Materials Required

100-cm (39-in) length of transparent plastic tubing, 3/16-inch inside diameter, 1/16-inch wall thickness (obtainable at many hardware stores)
Grain-of-wheat (pinlight) lamp, plain glass bulb, Archer 272-1140 (6 V), 272-1139 (1.1 V), or equivalent
LED, visible red
LED, green
LED, yellow
LED, infrared
CdS light-sensitive resistor, Archer 276-116 or equivalent
1.5-V cells or power supply
Black silicone sealant, GE or Dow type (obtainable at many hardware, auto supply, and paint stores)
Distilled water (obtainable at many grocery stores, or make your own)

Preparatory Remarks

The light pipe is made from a length of surgical plastic tubing with walls more or less transparent to invisible light. Your first series of experiments is made with a CdS cell, which is slow in reaction and less sensitive than silicon photocells, but the contrast in operation of the two types should be experienced to be appreciated. Each has its place in the work of electronics.

Procedure

Step 1

From a coil of 3/16-inch id (inside diameter) clear surgical plastic tubing, cut a 100-cm length, taking pains to make a smooth, perpendicular cut. A new razor blade or a sharp knife will make an excellent, rolling cut. You can save time while waiting for applied sealant to set by cutting four 2-cm pieces of the tubing from the supply coil, to be used for making end-plugs for sealing the LEDs into the liquid-filled tubing. See Fig. 7-1.

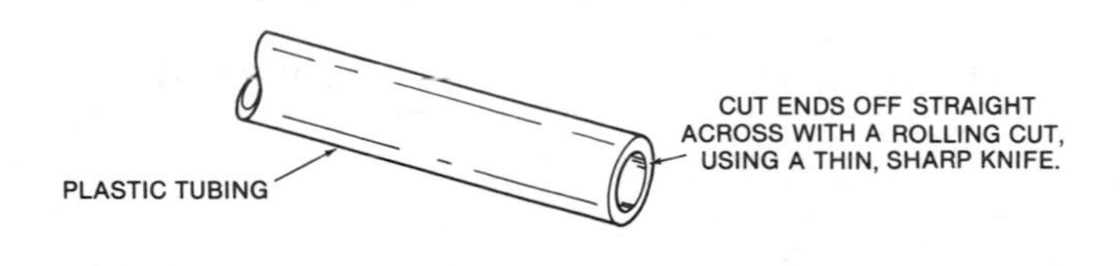

Fig. 7-1. Method of cutting plastic tubing.

Step 2

Check the operation and range of the CdS light-sensitive resistor by connecting it to a multimeter and using the ohms scales. With light shining on the face of the cell, note the reading on each of the ohms ranges. Then cover the face of the cell with a piece of electrician's black tape. Compare light and dark readings on each range, and note the multimeter range setting that gives the widest range between readings; this is the multimeter setting to use for your tests. Record your observations in Table 7-1. (Your meter may not have exactly the same ranges or all of the scales noted.)

Table 7-1. Resistance Readings for CdS Cell

Range	Dark Ohms	Light Ohms
100		
10 kΩ		
1 MΩ		

Step 3

Be sure that the lamp and LEDs are used at their correct operating voltages. If different voltages are required, fill in Table 7-2 for later reference to avoid destroying any items with overvoltage.

Table 7-2. Operating Voltages

LED or Lamp	Voltage Used

Step 4

Secure the CdS cell mechanically to one end of the tubing by any convenient means that will permit application of sealant without disturbing the face-to-face contact of the cell to the end of the tubing. One method is to use a removable pocket clip from a ball-point pen; the clip should just fit the outside diameter of the plastic tubing (Fig. 7-2). With long-nosed pliers, bend the long portion of the clip to lie alongside the light cell, and use rubber bands to hold it there. When the assembly is mechanically secure, cement the assembly by using a liberal amount of black silicone sealant to make a water- and light-tight seal between the cell and tubing. Any sealant on the cell face at the inner diameter of the "pipe" will prevent light from entering the cell, so care in the sealing operation is a must. See Fig. 7-3.

Step 5

Insert the grain-of-wheat ("pinlight") or other small incandescent lamp into the other end of the tubing, just barely beyond its own depth, leaving the leads sticking out. Press silicone sealant in behind it, plugging the tubing for about ½ cm (0.2 in). Be sure that the inner end of the bulb is free of any sealant, so that it will transmit its light down the tube. Lay the assembly aside to allow the sealant to cure, a matter of from two to six hours, according to the directions on the silicone-sealant tube.

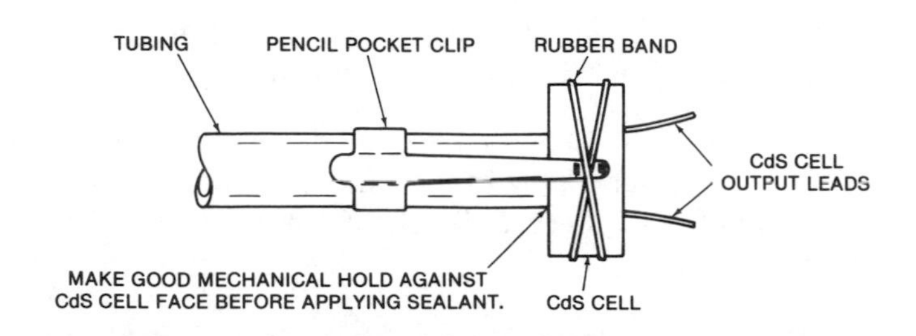

Fig. 7-2. Mechanical attachment of cell to tubing.

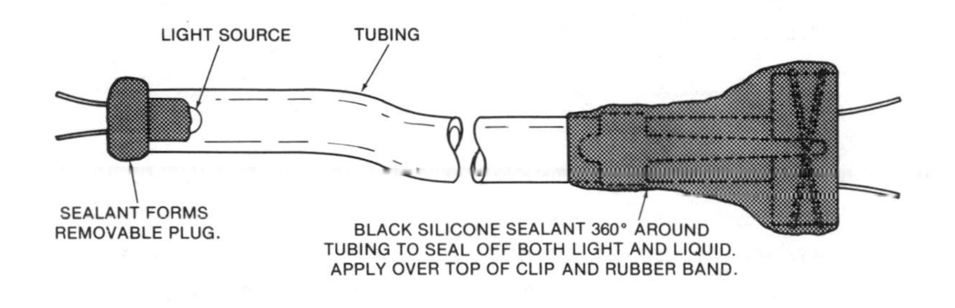

Fig. 7-3. Completed light pipe.

Step 6

Insert an LED in one end of each of the four 2-cm pieces of tubing called for in Step 1, and seal them there as with the lamp, with the leads sticking out. Set these aside to cure. When the sealant has cured to the consistency of rubber and is no longer tacky, you are ready for the next step in your series of tests. Cure time is dependent on the relative humidity of the atmosphere, because the catalyst is air-carried water.

Step 7

Connect the CdS cell to the multimeter as you did in Step 2. Note the reading in ambient light; it will be almost the maximum noted in Step

2, but this time the light is entering through the transparent wall of the tubing. Turn out your working lights and take another reading (with a shielded flashlight to illuminate the face of an older-type analog multimeter), or put a piece of black cloth or paper over your photocell and tube and take the reading. It should almost match the dark-photocell reading of Step 2, if it is not exactly the same.

With the lights out or the tubing covered with opaque material, connect the battery or power supply to the lamp in accordance with your voltage table from Step 3. Record the ohmmeter reading in Table 7-3, and note any change.

Table 7-3. Ohmmeter Readings for Step 7

Range	Reading

The resistance of the photocell may change slightly or not at all. Check to be sure that the lamp is not broken or burned out, and that the light enters the tube without being blocked by the silicone sealant.

Step 8

Remove the lamp assembly from the end of the tubing. You will find that the sealant acts as a ready-made "cork" that can be pried out without harm to the lamp. The sealant will not stick to the plastic tubing.

Using a suitable length of smaller-diameter tubing, fill the light pipe with distilled water. The smaller tubing may be used as a "pipette" by holding it in the supply jug of water long enough to take up water, and then placing your finger over the open end before lifting the tube out of the water. Insert the tube full of water all the way down into the light pipe, and release your finger. After a few of these operations, each time lowering the small tubing just below the surface of the water in the larger pipe, you will have your tubing filled to the brim without bubbles. If the water is poured into the end, bubbles will form that are very difficult to eliminate.

A bubble will interrupt the optical path, so be sure there is a solid column of water before taking the next step. Hold the water-filled tubing vertically at all times until the top end is plugged with the lamp/sealant assembly.

Note also that there is a difference between true distilled water and the so-called "distilled" water that actually is simply passed through two types of ion-exchange resins. If you find it difficult to obtain "real" distilled water, you might make your own by running steam from your teakettle through a length of tubing that is cooled by passing cold water over it.

Step 9

Lay a short length of thin copper wire (a strand from a piece of stranded hookup wire will do) alongside the plug containing the lamp as you insert it into the vertically held tubing. This will give the excess water an escape path as the plug is inserted; without such a path the pressure buildup could blow the CdS cell off the other end. Remove the wire when the cork is in place.

Step 10

Connect the lamp to a battery or power supply of the appropriate voltage, as recorded in Table 7-2.

Step 11

Take readings of the resistance of the CdS cell first without a cover over the tubing, and then after the tubing is covered with an opaque material to prevent ambient light from entering, as in Step 7. Record these readings in Table 7-4 for future reference.

Table 7-4. Resistance Readings for Step 11

Range	Resistance

Step 12

Remove the battery and meter connections, and spiral-wrap a strip of aluminum foil the full length of the tubing, with the shiny side facing the tubing. Common kitchen wrapping foil is quite usable. Scissors will cut the foil into strips better than a knife. Use tape to prevent uncoiling.

Step 13

Reconnect the multimeter and power source as before. Note the resistance of the CdS cell with the lamp on and the lamp off. Record the results in Table 7-5. The on-reading with water in the tube will have increased over the off-reading as recorded in Steps 7 and 11.

Table 7-5. Resistances for Step 13

Lamp	Resistance
On	
Off	

Step 14

Substitute the LEDs for the lamp one by one, taking care each time to make certain that the tubing is full of distilled water, free of bubbles. Use the wire alongside the plug to allow excess water to escape and to prevent build-up of pressure against the CdS-cell sealant on the

Table 7-6. Resistances for Step 14

LED Type	On Resistance	Off Resistance

other end of the tubing. Take readings with the LED on and then with the LED off, and record them in Table 7-6.

Step 15

Retain the water-filled tubing for the next experiment.

Commentary

There are at least two major variables to consider in evaluating your readings: the CdS cell has a spectrum of light sensitivity, and the LEDs may not all have equal light power output. It is possible that one or more of the LEDs will not show much of a resistance change from on to off because of these color and power characteristics. Another variable to consider is the water; it may not transmit all colors of light equally well.

You have probably found thus far that light travels better through distilled water than through air bounded by the plastic tubing. By testing each LED and the lamp placed directly on the CdS cell, it can be noted further that water attenuates light rather severely as compared to exposing the cell to the output of a light source at close range.

Another fact is that light will follow the contours of the water-filled tubing provided that any bend is not so sharp as to close off the tubing or severely constrict it.

The tubing was cut to minimum length for three reasons: The CdS cell is not very sensitive to light as compared to modern photodiodes and phototransistors. Secondly, water is not as efficient in conducting light as are some of the fiber materials now available. Lastly, the plastic tubing permits a considerable amount of light to escape through the water/plastic boundary, requiring the help of a continuous mirror (the aluminum-foil surface) at the second boundary. It is possible that the plastic itself is a better medium than water for the transmission of light — a few experiments might be worth while to test this for yourself.

The diameter of the water column is large compared with the wavelengths of light directed through it. This in itself can cause modal dispersion to be quite high. And even though relatively pure water is used, molecules of certain minerals and air remain among the water molecules, causing attenuation of the light waves. Thus, while the experiment was successful and interesting, a longer tubing would probably have little practical value as a light pipe in a project, although a phototransistor instead of a CdS cell might permit as much as 10 meters or more of water-filled tubing to be used. Further, larger-diameter "pipe" might carry more light for a longer distance, with the accompanying bulk penalty, of course. An enterprising experimenter may think of a truly practical use for the water-filled light pipe.

Certainly you have proved that a 100-cm length of water-filled tubing could be used to operate a relay and control small or large amounts of current as the presence or absence of light at the tube entrance is sensed by the photocell at the other end. Since the ohmmeter swing is appreciable with the application of light, the meter could be replaced by a suitable relay or amplifier. In a practical application, however, one would have to remember that water and winter temperatures do not mix well without antifreeze additives, and such additives may severely attenuate the light.

Would alcohol or other liquids be an improvement over water as a light conductor? It might be quite interesting to find out.

8 Experiment 2: Light-Beam Voice and Music Modulator

It is relatively easy to impress light upon a conducting medium and simply turn it off and on at a slow rate to convey information. It is a greater challenge to voice-modulate (impress information upon) a light beam, whether in air or through a conduit such as the water-filled tube. This experiment demonstrates amplitude modulation of the filament lamp. It uses materials that are readily obtainable, are quite inexpensive, and require very little preparation considering the surprising quality of the results.

Discussion

Pocket transistor radios can be obtained for a dollar each or less in the right places. People unfamiliar with electronics often use them until the batteries deteriorate and then throw them away, not realizing that a new battery will restore them to operation. If you don't have two or three lying around the house, you can find them in used-goods stores or at garage sales. Retail outlets sometimes have them on sale at very low prices.

One of these radios will be used to modulate the light beam and the other to "receive" the beam. The former must be operative as a radio (can be either am or fm), and the latter needs to be functional only as an audio amplifier (the two or three transistors following the detector stage, plus the speaker).

If a filament lamp of small size is connected to a radio in place of the speaker, the lamp will light brightly on loud passages of music or voice. However, upon trying to pick up the transmission with a photocell and amplifier, you would find a barely intelligible garble. This is because the lamp filament cannot heat up from cold to hot quickly enough to follow the rapid voltage-level changes of the voice signal. The filament reaction is slowed by *thermal hysteresis*; the filament heating lags the voltage peaks by as much as 180° or more. The lamp will work better on music of certain kinds because there is a greater average value of power involved, which tends to keep the filament partially lighted without permitting it to extinguish between passages.

By placing a dc bias on the lamp so that it burns steadily at about half brilliance with no voice or music present, the circuit acts much like a high-level amplitude modulator in a broadcast transmitter. The alternating-current signal will operate the filament from its half-hot point in both directions, cooling on negative excursions and heating on positive, never totally extinguishing. Turn-on, however, can exceed normal brilliance, introducing some distortion and endangering the filament. From the midpoint of its operating temperature, the lamp can more easily follow the variations in power coming from the signal.

The intensity-modulated light beam is received by the CdS cell, and the result is amplified by the second radio (modified to function solely as an audio amplifier). The CdS cell is relatively slow to follow the rapid fluctuations in intensity of light, so the results are interesting and instructive, but hardly practical. Replacing the CdS cell with a silicon photodiode will demonstrate an improvement in response to voice modulation. Replacing the filament lamp with a LED brings about an even greater improvement in response.

This experiment makes further use of the plastic water-filled tubing prepared for Experiment 1.

Materials Required

1 Plastic tubing from Experiment 1
2 Pocket transistor radios (am or fm)
2 9-volt batteries (and single cells if needed in either radio)
1 1.5-V C-cell

1 Battery holder for C-cell
1 1.5-V "grain-of-wheat" lamp, filament type, Archer 272-1139 or equivalent. (Higher-voltage types may be substituted, to disadvantage.)
1 10 kΩ resistor
1 Ceramic capacitor, 0.01 to 0.047 μF
1 Photodiode
1 Phototransistor, Archer 276-130 or equivalent
1 Transistor socket for phototransistor
Stranded, colored hookup wire
Distilled water
1 CdS cell (from Experiment 1)
1 LED (from Experiment 1; see text)
1 Clip lead
2 Output transformers (optional; see text)

Preparatory Remarks

The "transmitter" of the light-beam modulator, one of the transistor pocket radios, is modified quite simply by disconnecting the speaker leads. The second radio requires a slightly more extensive modification to make it serve solely as an audio amplifier. Any operating audio amplifier may be used if desired, including one channel of your hi-fi music system.

Procedure

Step 1

Mark the radio you have selected as your transmitter with a distinctive press-on label. Remove its back to gain access to the speaker connections, and, if necessary, remove the printed circuit board to get to these connections. Desolder or clip the connections at the speaker terminals, and extend the leads through a hole in the side of the plastic case. If the leads are too short, extend them with short lengths of hookup wire, and insulate the splices with tape or shrink-tubing.

Step 2

Connect the grain-of-wheat lamp to the two leads emerging from the radio case (Fig. 8-1). Turn the radio on and set the volume about half-way. Tune across the broadcast band, and watch for the lamp to light. Select the station that seems to do the best job of lighting the lamp. Turn the radio off, and set this portion of the experiment aside for the time being.

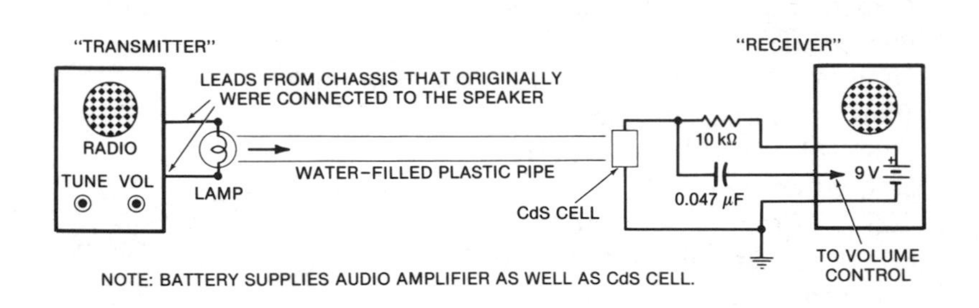

Fig. 8-1. Connections for lamp – CdS-cell setup.

Step 3

Remove the back and pc board from the second transistor radio, but leave the speaker leads intact. Solder a 21-cm (8-in) length of hookup wire to the pc board at a convenient point that traces through the power switch to the battery. The switch may connect to either pole of the battery, depending on the manufacturer. See Fig. 8-2.

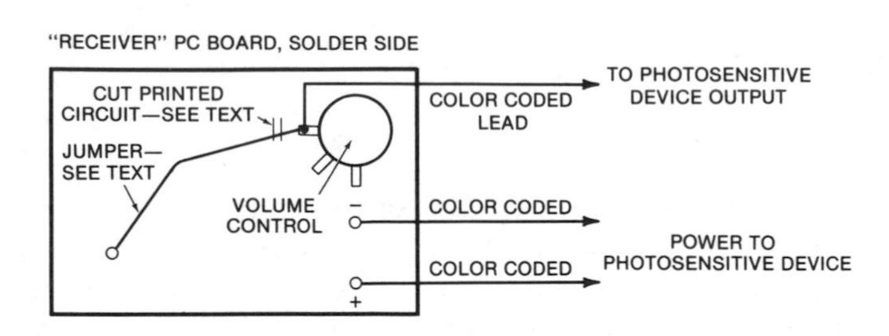

Fig. 8-2. Modifications to transistor radio to make light-pipe receiver.

Step 4

If the original audio-signal lead at the volume control is connected to the radio circuit by a jumper wire, simply remove the jumper and substitute a 21-cm (8-in) wire leading out from the volume control. If the volume control is soldered directly into one of the printed-circuit tracings, carefully cut a gap in the signal-lead tracing with a pocket knife or other sharp instrument. Make the cut as close as possible to the signal-input lead of the volume control (the position at which the wiper arm of the control would be when the control is turned to maximum volume). Record the lead color code directly on your schematic (Fig. 8-2) for later reference.

Step 5

Take another lead that is the same length as the other two leads you prepared, and solder it to the printed circuit board at any convenient point that traces to the pole of the battery opposite to that mentioned in Step 3. Be sure this lead is color coded for differentiation from the others, and note it on your schematic (Fig. 8-2). If the three wires are small enough in diameter, all three may be fed through the headphone jack, and the pc board can be restored to its mounting. Otherwise, drill a small hole in the case for the leads to pass through. Replace the back on the "amplifier," which now is actually your "light receiver" when connected to the CdS cell. (If you want to mount your CdS cell on a small board, note Fig. 8-8.)

Step 6

Insert the incandescent lamp into the water-filled tube as instructed in Experiment 1. Connect the CdS cell at the other end of the tubing to the receiver as shown in Fig. 8-1. It is assumed that you have retained the aluminum-foil wrapping around the water-filled tubing.

Step 7

Turn the transmitter and receiver on, and tune across the broadcast band with the transmitter. Compare the intelligibility of the resulting music and voice.

Discussion

As previously noted, there are a number of reasons why this particular setup will not perform too well. The transmitter must light the lamp to average brilliance by the power of the signal, which is impractical for voice but can be barely satisfactory for music. The CdS cell is relatively slow in response, and both voice and music suffer in quality as a result.

Continue the experiment as follows to improve the sound of both voice and music through your "light-pipe" system.

Step 8

Insert a C-cell battery holder in the lamp circuit of the transmitter, as shown in Figs. 8-3, 8-4, and 8-5. Place the C-cell in the holder and observe the lamp; it should light to about half normal brilliance, perhaps slightly brighter, without a signal. If the radio "transmitter" does not use an output transformer (some brands do not), connect two miniature output transformers back-to-back as shown in Fig. 8-5, with the C-cell in the circuit of the second transformer. Make certain that the higher-resistance windings are connected together rather than the low-resistance (usually 8-ohm) windings.

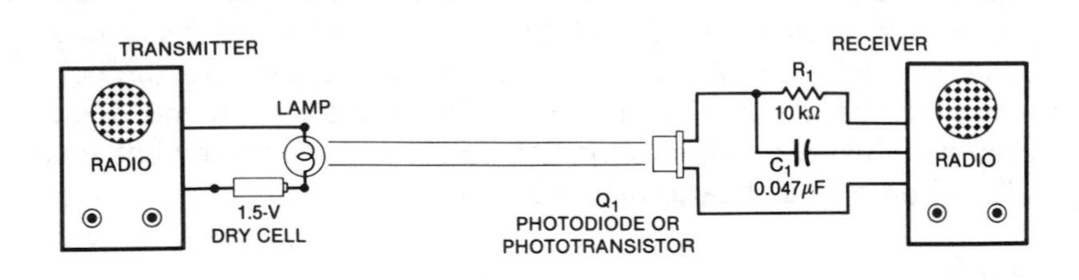

Fig. 8-3. Diagram of light-pipe communicator with biased lamp.

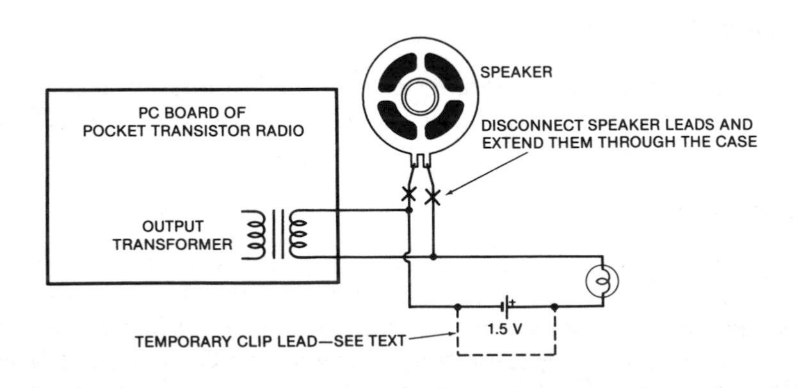

Fig. 8-4. Modifications to make transistor radio into light-beam modulator.

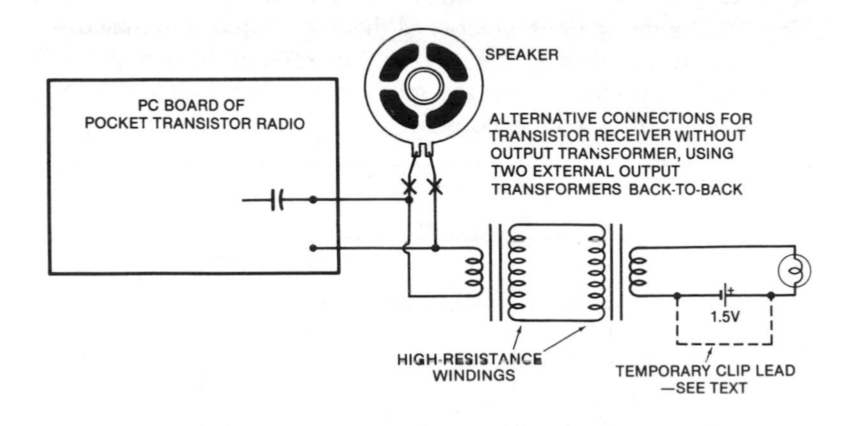

Fig. 8-5. Alternative "transmitter" connections for radio that does not have output transformer.

Step 9

Tune across the broadcast band with the transmitter, and observe the changes in brilliance of the lamp. The volume control should be advanced cautiously this time, to avoid burning out the bulb with a strong signal. The lamp shines more brightly all the time, although the fluctuations in light level are now much less pronounced than in Step 2. On the surface, it would seem that the volume and signal quality would be reduced, but perform the next step and see what happens.

Step 10

Turn the receiver on and tune across the broadcast band with the transmitter. An improvement in intelligibility is quite evident, is it not? Why is this the case?

This happens because the lamp is preheated by the bias current and does not need to reheat from cold each time the amplitude of the signal increases from a zero-point during modulation.

Step 11

Prepare a photodiode in a manner similar to that used for the CdS cell. Use silicone sealant to make the plug watertight, and at the same time be careful not to get the sealant between the open bore of the tubing and the transparent window of the photodiode. You may use a phototransistor in this step if you wish, but keep in mind that it will require less amplification from the receiver than the photodiode. See Figs. 8-6 and 8-7 for connections.

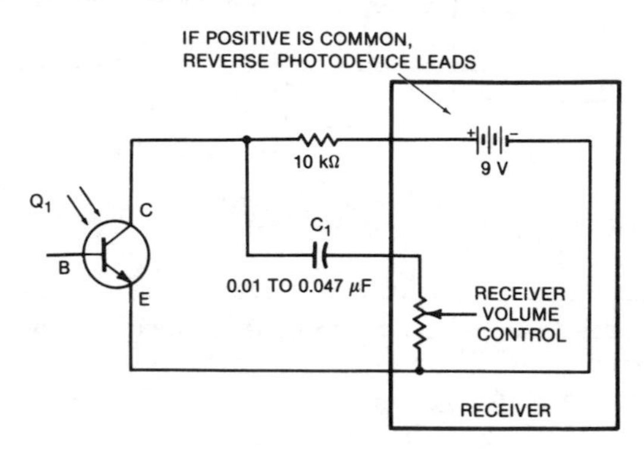

Fig. 8-6. Connection of photodevice to radio that uses 9-volt battery.

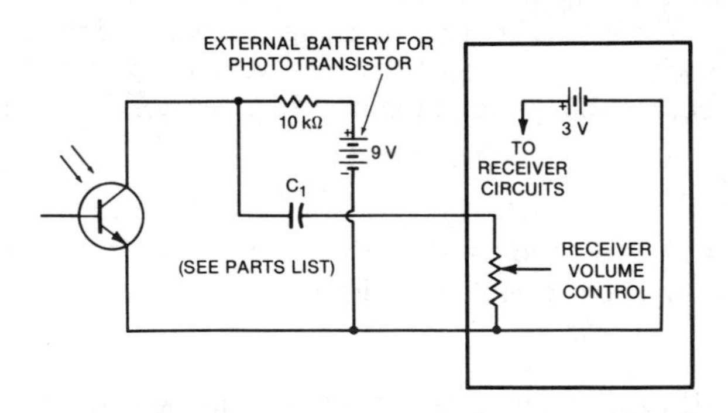

Fig. 8-7. Connection of photodevice to radio that uses one- or two-cell battery.

NOTE: Some photodiodes have their maximum sensitivity to light perpendicular to the leads, but others are manufactured to accept light end-on. Be sure which type you have, and insert it appropriately in a short bit of tubing; then seal it with the silicone compound. After the sealant has hardened, you may transfer it to the water-filled tubing. The "window" of the photodevice must face the water column and be free of sealant.

Step 12

When the sealant has cured, remove the plug from the short bit of tubing, and insert it in the water-filled tubing, again making sure that there are no bubbles of air introduced. Connect the photodevice in the bias circuits in place of the CdS cell, as in Figs. 8-6 and 8-8.

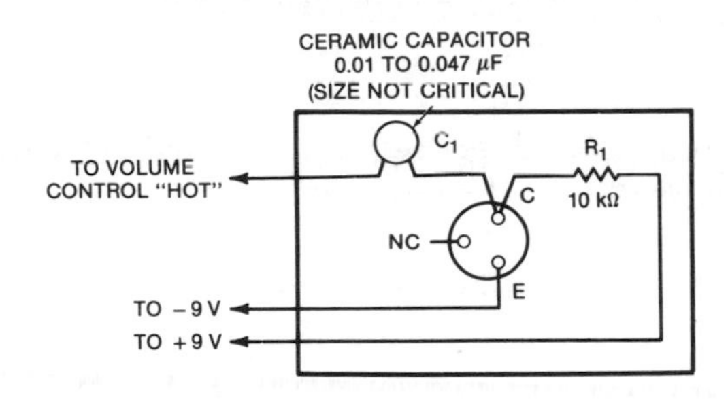

Fig. 8-8. Layout for mounting of photodiode or phototransistor and related components on perforated board.

NOTE: A phototransistor, as opposed to a photodiode, will require a full nine volts, but will be protected with a series resistor as noted in Step 14.

Step 13

Again listen to the transmission of music and voice through your light pipe, and note the improvement in intelligibility. Why is there an improvement? The photodiode or phototransistor is far more responsive to high-frequency light changes than the CdS cell.

Discussion

Considerable improvement in transmission quality over that experienced heretofore will now be apparent. Both transmission and reception are enhanced, and yet there are two additional improvements still to be made. You may well be surprised at how well water is performing as a light-conducting medium.

Step 14

A phototransistor will now be substituted for the photodiode at the receiver, if you have not done so already, using the schematic in Fig. 8-6 or Fig. 8-7. You may wish to assemble these components on a breadboard or perforated board (Fig. 8-8). Observe the polarity of the photodevice, as determined by trial.

If the radio in the receiver of your system uses two penlight cells, you must use an external 9-volt battery to energize your phototransistor circuit. In this case, only the common, or "ground," lead and the signal lead are connected to the receiver (Fig. 8-7).

Step 15

Turn on the transmitter and receiver, and note the signal quality and strength as compared with the results in Steps 7, 10, and 13.

Step 16

Select the LED determined in Experiment 1 to be the best for operation through your light pipe, and put it in the transmitter circuit instead of the filament lamp. Remove the C-cell from the bias-cell holder, and short the holder contacts together with a clip lead for the first test (Figs. 8-4 and 8-5).

Step 17

Turn on the receiver and transmitter, and note the transmission quality as compared to the previous attempts. Remove the clip lead from the battery holder.

Step 18

Make certain that the battery polarity is correct, and insert the bias cell in the LED circuit. Turn on the receiver and transmitter and note the transmission quality.

Commentary

With Step 18 you have the final setup in this experiment, and the last of the experiments using the water-filled tubing. The final step will have slightly improved listening quality, although the difference may not be readily noticeable. Quality is, on the whole, remarkably good, considering the large diameter of the tubing and the potential mismatch between components.

You have seen, however, that amplitude-modulated light transmission through water for certain distances is quite feasible.

Certain practical uses for the system may suggest themselves to you, although the limitations may rule against any permanent installation. Hold off making any use of the water light pipe until you have performed the experiments that follow, which involve glass fibers as the light-conducting medium. The vast improvement in effective distance and much improved quality of transmission should make the glass fiber a "better buy" in virtually any practical application.

Retain your "transmitter" and "receiver" for the next experiment, but you may retire your plastic tubing to more mundane uses. The hands-on experience of the preceding two experiments will have given you a memorable feel for the potential inherent in communicating on a light beam. You never know when such practical knowledge will prove to be invaluable in, perhaps, an emergency.

Can you think of some uses for fiber-optics devices, now that you have had some practical experience with them? What about the potential safety of a light fiber? Would it be possible to run a light-carrying fiber through an explosive atmosphere without danger of igniting the gases? How about fiber-optics communications in a nuclear submarine or land-based power station? In case of an enemy nuclear attack, would there be an advantage to having fibers instead of copper wires in buried conduits along the highways?

9 Experiment 3: Improved Modulator

Using the simple home-workshop transmitter and receiver built for the previous experiments, you will prepare and insert an optical-fiber strand or cable between the light source and the detector. You will be able to evaluate whether the fiber improves the quality and distance of communication enough to warrant its use.

Discussion

Fiber-optics cables or single-strand fibers are not as easy to obtain for experimentation as wire. A few supply houses dealing in factory-overrun merchandise occasionally advertise various lengths and types of fibers, but they do not repeatedly stock any given item. Edmund Scientific has a small selection currently available (see Appendix D).

A few companies will answer inquiries with a sample of their product. At this time, ingenuity and persistence are the best tools to employ in your search for fiber. A number of companies have supplied information at our request, and a few have supplied samples for the experiments and projects in this book. They are listed in Appendix D. Manufacturers in general sell only through distributors or agents, and they will supply names of these upon mail or phone inquiry.

The initial part of the experiment will use the same small filament lamp, photodiode, and phototransistor used in Experiment 2, although you may wish to leave the former setups intact and opt for new parts for this one. There is no need to use the CdS cell, as the former experiment established that it is too sluggish to follow the rapid fluctuations in light level required for good voice communication.

Materials Required

Modified transistor radios from Experiment 2
Battery holder
1.5-V filament lamp, Archer 272-1139 or equivalent
Photodiode, Archer 276-144 or equivalent
Phototransistor, Archer 276-130 or equivalent
Fiber-optic strand or multistrand cable
Fiber-optic polishing kit or a selection of three to six fine grades of emery dust
Miscellaneous insulating spaghetti and heat-shrink tubing

Procedure

Step 1

The type of fiber strand or cable that you have obtained will affect to some extent your approach to preparation of the ends. If you have obtained the AMP or other experimental kit, instructions accompanying it should be followed in detail.

Assuming that you have acquired a multiple-strand cable such as Belden 200, gather a miscellaneous assortment of bits of spaghetti insulation, including bits of plastic insulation from electrical wires. Assorted sizes of heat-shrink tubing are particularly useful, because they will draw tight with the application of heat for a secure assembly.

Step 2

The 1.5-V filamentary lamp, although quite small, is much larger than the fiber-optics cable. It is gigantic compared to a single strand of fiber. Select a piece of spaghetti that fits firmly over the bulb, but use care as you slip it over the bulb because the bulb glass is quite thin and brittle.

Step 3

Build up the cable by wrapping a half-inch strip of thin paper around it at the end until the paper matches the bulb diameter. Follow the sequence in Fig. 9-1:

A. Wrap the strip tightly in a spiral around the fiber.
B. Estimate the diameter needed. It is easier to remove a few turns than to add turns to an ill-fitting job.
C. Slip a piece of heat-shrink tubing over the completed wrapping.
D. Apply heat evenly all the way around the assembly. Excess diameter can be reduced with emery cloth, but be careful not to scratch the polished end of the fiber.

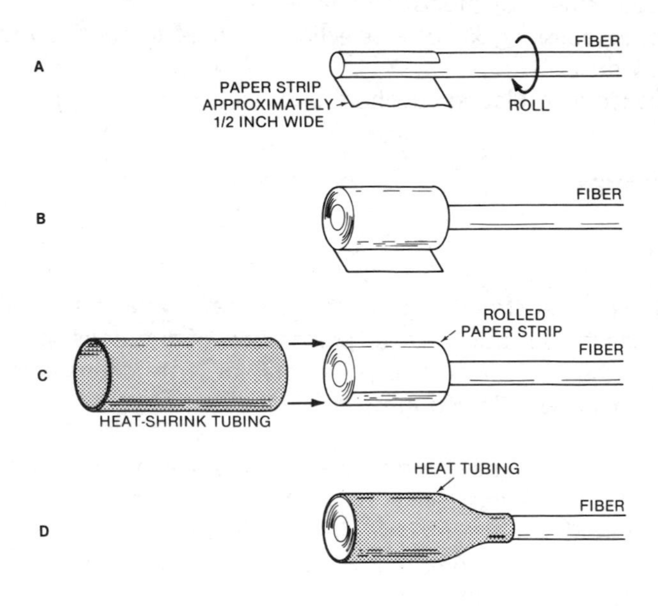

Fig. 9-1. Preparation of fiber end for experimental coupling.

Step 4

If the cable has not been factory cut and polished, carefully grind and polish the fiber end against four to six pieces of increasingly fine emery paper or dust, from 200 grit to finer, as set forth in Appendix F. Finish with a fine grit-based polishing compound, such as that sold for copper or silverware. Some types of toothpaste work well for polishing, while others contain no grit. Hold the assembly perpendicular to the emery paper, and use a circular motion to polish. The emery paper or dust should be laid out on a hard, flat surface for grinding.

Step 5

Slip the cable with its built-up end into the piece of spaghetti that was fitted over the bulb in Step 2. Shrink tubing will work very well to make a snug fit (Fig. 9-2). Make sure the cable butts against the end of the lamp bulb, aligned as close to the filament as possible, before cementing or applying hot air to shrink the tubing. (If you use one of the electric hair dryers for this task, or a hot-air gun made for the purpose, rotate the assembly as you apply heat.) Instant cement such as "Crazy Glue" can be used to bond other types of spaghetti securely, provided it is tightly fitted.

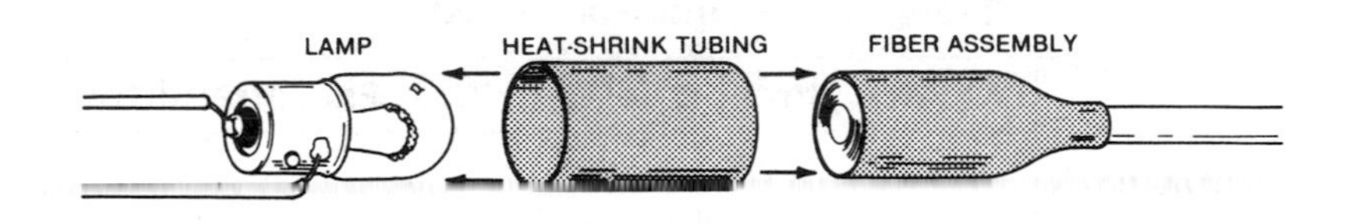

Fig. 9-2. Method of fastening fiber assembly to lamp.

CAUTION/WARNING: Instant cement is potentially hazardous to the eyes. If you have ever gotten it on your fingers and found them cemented together so tightly that you would lose a layer of skin to pull them apart, you can imagine what would happen if you were to splash any of it in your eye. It is aptly named: it reacts instantly, before you can blink or pull your fingers apart, and the closer the fit between the parts to be cemented, the more quickly it reacts. Use only with great caution, and DO NOT LEAVE IT WITHIN THE REACH OF CHILDREN! Acetone or some types of nail-polish remover may be used as a solvent to remove the set glue from your fingers, but they are NOT to be used for sensitive tissue such as the eyes.

Step 6

Prepare the opposite end similarly, but mount the photodiode in place of the lamp. If you have only a very limited supply of fiber or cable, you may want to prepare this and the light-source end in such a way that the fiber can be inserted and removed without cutting or breaking.

Step 7

Attach the bulb to the transmitting circuit used in Experiment 2 and the photodiode to the receiving circuit. Note the volume and quality of voice and music transmitted.

Step 8

Carefully slit the heat-shrink tubing, or use other means of your own devising, and replace the photodiode with the phototransistor. Compare your observations of the results in Table 9-1. Chances are you will be unable to detect any quality improvement, but a twofold or threefold volume increase is likely.

Table 9-1. Comparison of Results

	With Photodiode	With Phototransistor
Volume		
Quality		

Commentary

You will most likely have to reduce the volume of both transmitter and receiver when the fiber cable is used. Compared with the water-filled tubing, the glass fiber is much more efficient for transmitting light. This is most noticeable when the phototransistor is used in the experiment, since this device is not only sensitive to light but also amplifies its output current. In addition, the quality of the throughput should have improved over that of the water medium.

Photodetectors have a sensitive spot that is centered only approximately in the less expensive products. Precision products, such as those sold by Motorola for the AMP experimenter's kit and by Sylvania for industrial replacement purposes, have sensitive spots precisely coaxial with the outer body of the unit.

Although the less expensive unit is only approximately centered, a cable containing upward of 200 fibers will not be difficult to place effectively. But, if a single strand of plastic or glass fiber is used, placement is highly critical, and the construction will have to be very carefully done. Be aware that the single strand may have to be placed slightly off-center relative to the physical outline of the device to accommodate the lack of precision in solid-state manufacture. You may wish to delay permanent sealing of the fiber to the photosensitive device until you have exactly located the "hot spot" by trial and error while your transmitter and receiver are operating.

Retain the transmitter, receiver, and cable for the next experiment.

10 Experiment 4: LED-Driven Fiber-Optic System

The purpose of this experiment is to demonstrate a competent, working system approaching industrial quality.

Discussion

While you may be able to use the fiber cable or strand from Experiment 3 for this experiment, it is preferable to take some pains to determine that your LED and fiber are compatible as to wavelength. The wavelength of the light output from the LED should match very closely the transmission specification for the fiber. Manufacturer's specification sheets are the best source for this information, although cut-and-try methods can be useful in their absence.

The assorted packet of LEDs available from Technical Electronics Corp., Catalog No. 1501 (see Appendix D), may be used for trying many different LEDs in hopes of finding a compatible LED for your cable. Radio Shack sells an infrared pair, LED and phototransistor, which might well save you hours of extra work, provided your fiber cable will transmit infrared radiation.

Note in Fig. 10-1 that we have added a potentiometer to the basic circuit of Experiment 2. This permits selecting the best bias level for quality transmission. (If the radio used does not have an output transformer, external transformers will have to be added in the manner of Fig. 8-5.)

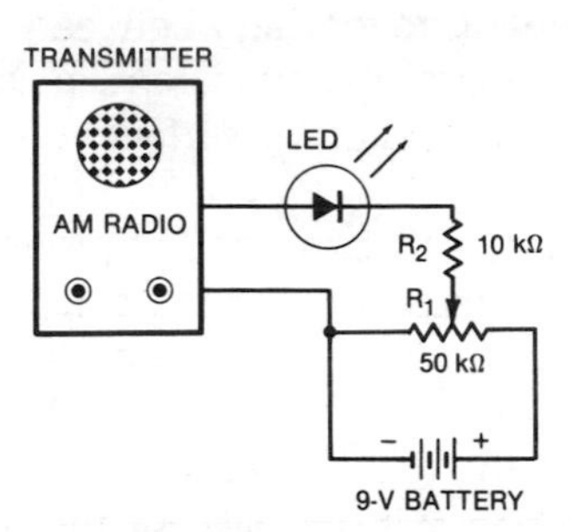

Fig. 10-1. Schematic diagram of bias-adjusting potentiometer and current-limiting resistor for LED.

Materials Required

Resistor, 10 kΩ
Potentiometer, 50 kΩ
Fiber strand or cable
Fiber polishing kit or selected materials
Assorted LEDs
Phototransistor
Previously prepared electronics from Experiments 2 and 3.

Procedure

Step 1

Prepare the cable ends very carefully if they are not factory cut and polished, by fitting them into a suitable holder or manufactured coupling. The procedures in Experiment 3 may be repeated here, but with even greater attention to detail. Grinding should follow the directions given in Appendix F.

Step 2

If the end of the LED is rounded to act as a lens, grind it flat, coming as close to the chip inside as possible without actually touching it or the leads emerging from it. Polish the flattened portion by the same method used with the fiber ends to obtain a high surface polish. While plastic is softer than glass and grinds more quickly, it is easier to scratch in the final polishing steps. Therefore, it requires even greater attention to washing between successively finer grades of emery than does glass. A transparent, polished surface transmits light far better than a translucent surface, on either type of fiber.

An alternative method of preparing the LED is very cautiously to drill a hole directly toward the chip, using a drill barely larger than the fiber or cable. Do not penetrate the chip. The bottom of the hole may be polished with a very smooth, flat-ended round toothpick no larger than the drill, with emery dust as the polishing compound. Wash out the hole with distilled water, and let it dry.

The polished end of the fiber may be inserted into the hole and transparent quick-set epoxy used to cement it there permanently. Instant glue may likewise be used. Again, care and patience are the watchwords. How well you prepare the emitter and detector in great measure affects the quality of your transmission.

Step 3

Mount a phototransistor on the other end of the cable or fiber strand, using a method similar to that described above, as determined by your components and skills. Assuming that your cable will be a short run (under 3 meters, or approximately 9 feet), liquid wetting of the gaps between the light-generating and light-detecting elements and the fiber will not be necessary.

Step 4

Connect the components to your transmitter and receiver from Experiments 2 and 3, or better still, between a high-fidelity preamplifier and its amplifier. If you choose the second course, take care to keep the preamplifier volume control at a low setting until you can be certain that the output does not damage the LED by overdriving it.

Step 5

Compare the voice and music quality obtained through your fiber-optic system with the same voice and music routed through your components with direct electronic connections by wire.

Commentary

The quality of output from your transmitter and receiver as a fiber-optics system should be indistinguishable from that obtained with the transmitter and receiver connected by wire. Depending on the accuracy of match of your fiber-optic components, system quality may have differences detectable by such equipment as a distortion analyzer.

If the above proves true for you, you have built a fine, potentially useful analog communications setup, which may be retained in its present form for demonstration purposes or other practical use. A more formal unit of more attractive appearance is proposed for construction in the projects section of this book.

This concludes the series of experiments for demonstrating analog fiber-optic communications. Next, we tackle pulse transmission and reception.

11 Experiment 5: Basic Digital Transmission of Data

The purpose of this experiment is to demonstrate a simple method of transmitting data in digital form from one point to another, via fiber optics.

Discussion

While the filament lamp can be used to convey audio data from dc to about 5 kHz via fiber optics, its ability to be modulated at higher rates is rather limited, as we have demonstrated in previous experiments. Because the LED turns on and off rapidly, it can be used for the generation of short and fast light pulses.

An outstanding advantage of digital transmission of data is that the pulse heights (voltage levels) can be made identical, pulse for pulse. Extraneous noises represented by variations in voltage, as encountered in amplitude modulation, do not affect digital pulses. A fiber will carry digital pulses as readily as amplitude-modulated light.

If a pulse reaches the end of its travel somewhat ragged on top, it can be restored to its original shape by amplifying and clipping, and retransmitted for another long optical run. The modulation remains intact, since it appears in the form of varying pulse widths or numbers of pulses per unit time, depending on the method of modulation that is being used.

Materials Required

1 555 timer chip, Archer 276-1723 or equivalent
1 50-kΩ resistor
1 10-kΩ resistor
1 1-kΩ resistor
1 100-ohm resistor
1 0.05-μF ceramic capacitor
1 0.1-μF Mylar capacitor
1 9-V battery
1 Red LED
1 Experimental board, Archer 276-174 or equivalent
1 Chip insertion tool, Calectro 9481 or equivalent
1 Selected photodiode or phototransistor (see text)
1 Telegraph key or convenient off/on switch
Receiver setup from preceding experiments

Preparatory Remarks

The experimental board makes for easy changing of parts and parameters in an experiment, since wires and components plug into holes that contain pins connected in certain prearranged configurations to other pins on the board. Although not absolutely necessary to successful performance of this experiment, such a board is so handy that it is highly recommended. Further, the manufacturer makes a pc board with pre-etched patterns exactly duplicating those of the experiment board, for easy permanent construction of a circuit that has been found to work. You simply remove the components from the experiment board and insert them in the corresponding locations on the pc board. The circuit is preconnected except for perhaps a few jumper wires.

Procedure

Step 1

Insert the chip in the experiment board with the insertion tool, in a position selected to make its connecting pins readily available to parts that will be inserted around it. A near-center position is suggested.

Step 2

Wire the chip with its resistors and capacitor according to the schematic diagram shown in Fig. 11-1. A new, selected LED, photodiode, and phototransistor are recommended, rather than trying to salvage those used in previous experiments. Note the dual listing of values for R_1 and C_1. Use the 50-kΩ and 0.1-μF values first.

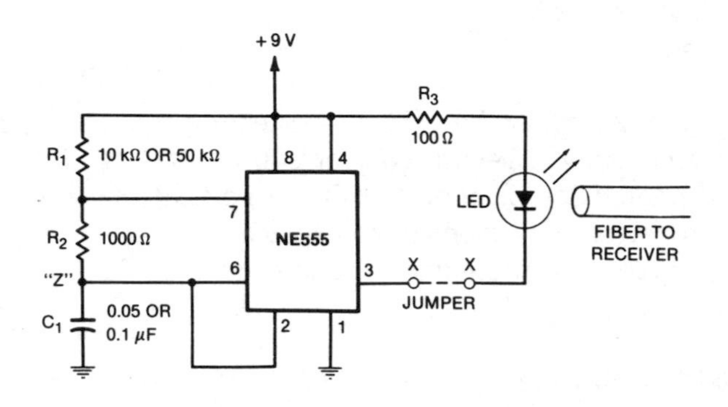

Fig. 11-1. Circuit for pulsing an LED at about 10 kHz.

Step 3

Connect the battery, observing polarity. Does the LED light? If so, go to Step 4. Single-color LEDs are polarized and will not light if inserted into the circuit incorrectly. (Some dual-color LEDs shine red or green depending on which polarity is applied.) If your LED fails to light, reverse its leads in the board and try again. If it still fails to light, try another LED, and check your circuit. Bring the receiver photodiode or phototransistor up close to the LED.

Step 4

Turn the receiver on; increase the volume until you hear a high-pitched tone, indicating that the LED is pulsing at about 5 kHz.

Step 5

Change R_1 and C_1 to the smaller "inside" values, and listen once more for the tone. This time it should be about 10 kHz.

Step 6

Open the circuit at "X . . . X" in Fig. 11-1, and insert the telegraph key, a switch, or a push button. Operate the key and listen to the result in the receiver. You are starting and stopping (modulating) strings of identical pulses at a rate determined by the length of time that the switch or key is closed. In this way, you could send International Morse code or some other on-off code.

Step 7

Disconnect the telegraph key and reconnect the LED directly to Pin 3 of the chip, in preparation for the next experiment.

Commentary

Modulation in this case is a simple on/off or H(igh)/L(ow) situation. If you know the Morse code, or are interested in learning it from Table H-2 (Appendix H), you can use this setup to communicate by light fiber over any practical distance with no interference from static, cross talk, or other signals. Two fibers side by side would permit two-way communication (plus, of course, another transmitter and another receiver until you become familiar with multiplexing and "break-in" operation such as amateur radio operators use). With multiplexing, a single fiber for communication in both directions is possible.

Each pulse is like the one before it. Each time you press the key for a "dit," you might send 100 pulses before your released key stops the string. Each time you press the key for a "dah," you might send 300 pulses. Of course, depending on the duration of your particular dits and dahs, you may be sending 1000 and 3000 pulses, or even 10,000 and 30,000. The number is not important, as long as there are at least two for a dit and six for a dah (you would need a computer to read Morse code at such a high speed). The higher the speed employed, the fewer pulses there will be in each dit and dah, if the chip-generated frequency remains fixed.

12 Experiment 6: Multitone Modulation of Digital Transmission

This experiment will demonstrate a second type of modulation for a fiber-optics pulse (digital) transmission system.

Discussion

One of the common methods of operating teletypewriter machines from a remote transmitter location is to encode the intelligence to be sent in the Baudot or ASCII code and send the result "serially." This means, in effect, that each typewriter key has a different code assigned to it, and the code for each letter or number consists of a number of off and on conditions. As each code series reaches the receiver, the typewriter will advance one space and print a corresponding character symbol. This is called the loop system, suitable for wire hookup. (The ASCII code is presented in Table H-1, Appendix H, for your information.)

In radioteletypewriter circuits, two major problems encountered with the above system were static (atmospheric interference) and adjacent channel interference. By using the Baudot or ASCII code but expressing the code with two tones, one tone for off and the other for on, most of the interference could be handled without excessive garbling of the received message. Filters in the receiver make each tone distinctive and tune out the interference. In this experiment, we discover that a fiber-optic system will work with a two-tone code.

Materials Required

1 0.01-μF ceramic or mini-Mylar capacitor
Transmitter and receiver setup as used in Experiment 5

Procedure

Step 1

Make sure the wire between "X . . . X" (Fig. 11-1) has been replaced and that the LED lights when the battery is connected. Review Experiment 5 for the detailed setup and checkout steps.

Step 2

Connect the telegraph key or switch in series with C_2, a 0.01-μF capacitor (Fig. 12-1). Connect this combination across C_1 (Fig. 11-1).

Step 3

Connect the battery, and listen to the basic tone. Press the key and listen to the lower tone. Then rapidly depress the key and release it, producing a series of warbling high and low tones. This is very nearly the sound of radioteletypewriter signals that can be picked up on a shortwave receiver that is equipped with a bfo, or beat-frequency oscillator.

Step 4

Remove C_1, C_2, and the telegraph key from the circuit in preparation for the next experiment. Leave all other connections intact.

Commentary

This same technique is used when computers and terminals transmit information to each other; the modulating-demodulating device is called a *modem.* Can you suggest some advantages and disadvantages of using fiber-optic cables in place of wire cables?

We have now demonstrated a basic form of pulse-frequency modulation (pfm). Each pulse is the same height (voltage value) regardless of

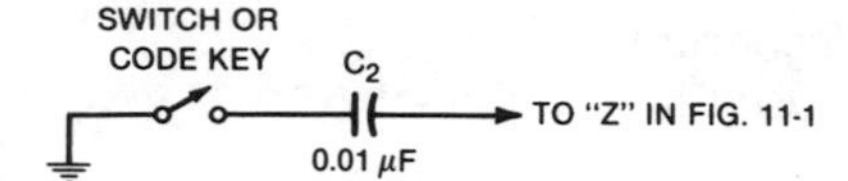

Fig. 12-1. Modification of circuit for Experiment 5 to make warble-tone modulator.

tone. In practice, each tone would be selectively accepted by a filter tuned to its frequency, with all other tones excluded. Note that these pulses travel through the fiber in the form of light, as readily as through a copper wire—and with even better fidelity because there are no inductive or electrostatic effects to cause distortion.

If you have teletypewriter equipment with which to transmit and receive, you may, obviously, operate your equipment with fiber optics as your transmission medium. In fact, the tiny transmitter and receiver you have just put together would serve nicely with only a few refinements.

Does this give you a "feel" for the remarkable simplicity of the fiber-optics system of communication?

13 Experiment 7: Basic Pulse-Frequency Modulation

The purpose of this experiment is to demonstrate a stepped-frequency modulation method for use with fiber-optics systems.

Discussion

There are thousands of uses for fiber optics that have yet to be developed. One of these might well be in building electronic organs, musical instruments of great tonal beauty and diversity of sound. One fiber might convey the results achieved in the mixing circuits to the amplifier, or thousands of individual fibers might be used to advantage in generation and distribution of these tones within the organ before amplification. The type of modulation produced in this experiment might lend itself to just such a project.

Pulse-frequency modulation, or pfm, may be generated by a simple arrangement based on a 555 timer chip. It can be made as elaborate as the experimenter desires, employing many capacitors in parallel or series-parallel to achieve just the right tone to suit the ear and the musical scale. One should not overlook the music synthesizer, of which the Moog instrument was the first (and probably most famous for its having been used to produce the composition "Switched-On Bach"). We will not attempt anything quite as elaborate, although our construction could have such a basic application. The materials list performs double duty as a table of resultant frequencies.

Materials Required

All items for Experiment 5, plus the following:

	Item	*Resultant Frequency*
C_1	500 pF ceramic	300 kHz
C_2	1000 pF ceramic	150 kHz
C_3	2000 pF ceramic	86 kHz
C_4	3300 pF ceramic	62 kHz
C_5	0.01 μF Mylar	24 kHz
C_6	0.02 μF Mylar	14 kHz
C_7	0.05 μF Mylar	9 kHz
C_8	0.1 μF Mylar	3.5 kHz
C_9	0.22 μF Mylar	1.5 kHz
C_{10}	0.47 μF Mylar	0.8 kHz

10 switches, push-button, no (normally open) or other convenient spst switches

NOTE: If only a few frequencies are to be tested, use capacitors from C_6 to C_{10} for audible sounds.

Procedure

Step 1

Connect the capacitors and switches to the basic 555 timer chip as shown in Fig. 13-1.

Step 2

Connect the battery, and listen to the result as each switch is pressed in turn. The complete fiber-optic cable assembly should be in use at this time.

Step 3

Press various combinations of switches, using all fingers and thumbs if possible. Note the tones produced and the clarity with which they are received through your fiber-optics system.

NOTE: In the first five switch positions, the frequencies produced will be inaudible, although detectable quite easily with a good oscilloscope.

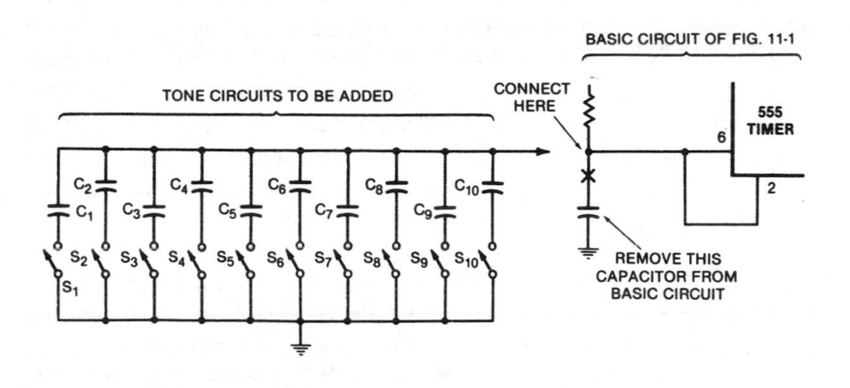

Fig. 13-1. Modification of circuit for Experiment 5 to produce stepped-tone modulation of pulse frequencies.

Step 4

After you read the following commentary, remove the step-capacitor circuit from the board in preparation for the next experiment. Leave all other items and circuits intact.

Commentary

The frequencies indicated in the combined materials list/table of frequencies are approximate, of course. They will vary depending on the tolerance of the capacitors, the battery voltage, and the tolerances of the resistors in the circuit. However, since the frequencies desired can be obtained by substituting other capacitors or combinations of capacitors, you may wish to "tune" your circuit to play the chromatic musical scale.

The frequencies listed are simply for indication of what should occur, and you will find that there is only a rushing sound or silence above a certain frequency. There is not enough space here to indicate the many variations in tone that are possible with this simple circuit, even if the upper frequencies generated are kept below the threshold of hearing.

Most important is your attention to the fiber-optics portion of your experiment, and the ideas you may generate as to how you might employ the fiber system in musical instruments or other practical ways. Note that the fiber system conducts even the highest "hearable" tones as easily as if they were transmitted by wire. If available for your use, an oscilloscope with twin-trace features will demonstrate the "hearable" pulses being received virtually as transmitted (scope input 1 at Pin 3 of the 555 IC, input 2 at the receiver output). The very high-frequency pulses will not get through the receiver audio circuits, but they may be measured at the receiver input.

From this setup, we need only introduce a variable capacitor or capacitor microphone in the circuit to make possible the production of pfm by voice. A similar arrangement could be used with an instrumentation transducer sensing some change in an observed parameter of a complicated machine under test. Does a form of security monitor suggest itself? There are 20- to 30-kHz transducers on many older tv consoles as part of the remote control system. Such televisions are to be had for the asking at radio shops, and for only a few dollars at Goodwill workshops. Might two of these transducers be connected into your experiment to make a combination ultrasound/fiber-optics security system?

Instead of multiple capacitors, the capacitance might be varied at a fixed rate, perhaps with a motorized capacitor or function generator; would your receiver be able to sense a Doppler shift in frequency under any external circumstances fed into the receiver in parallel with the LED light input? Might a redundant, obvious wire hookup be clipped by a burglar, thinking he had disabled the alarm while he overlooks a tiny, virtually invisible fiber-optics thread?

Although the range of capacitance required for modulating this particular experiment by voice would seem to be rather great, selecting a basic frequency high enough to require only a small capacitance change to cover the voice range would make voice modulation possible with this circuit. However, the question becomes one of reception: would the present receiver respond to pfm? Or is a special detector required? We will find out in the next experiment.

14 Experiment 8: Pulse-Frequency Modulation

The purpose of this experiment is to demonstrate successful transmission and reception of digital pulse-frequency voice modulation over a fiber-optic link.

Discussion

If we vary a constant-level, high-frequency square wave, causing it to increase and decrease in frequency from a median value at a rate of change determined by the human voice, we have achieved pulse-frequency modulation, or pfm. The amount of deviation from the center frequency will be determined by the amplitude of the voice, while the number of deviations per second will correspond to the voice frequencies. To modulate by pfm, we must set the center frequency to at least twice the highest frequency in the intelligence to be carried.

There are a number of possible ways of causing a 555 chip connected as a square-wave generator to vary in pulse frequency. Either C_2 or R_1 in Fig. 14-1 could be varied at the modulating rate. Of the two, the latter is easiest to accomplish for our purposes. If we connect the frequency-determining resistor, R_1, in parallel with the secondary of a microphone transformer, the combination will appear as a varying resistance to the chip. Capacitor C_1 prevents the low dc resistance of the transformer secondary from shorting R_1 almost completely, while passing the ac audio signal quite readily.

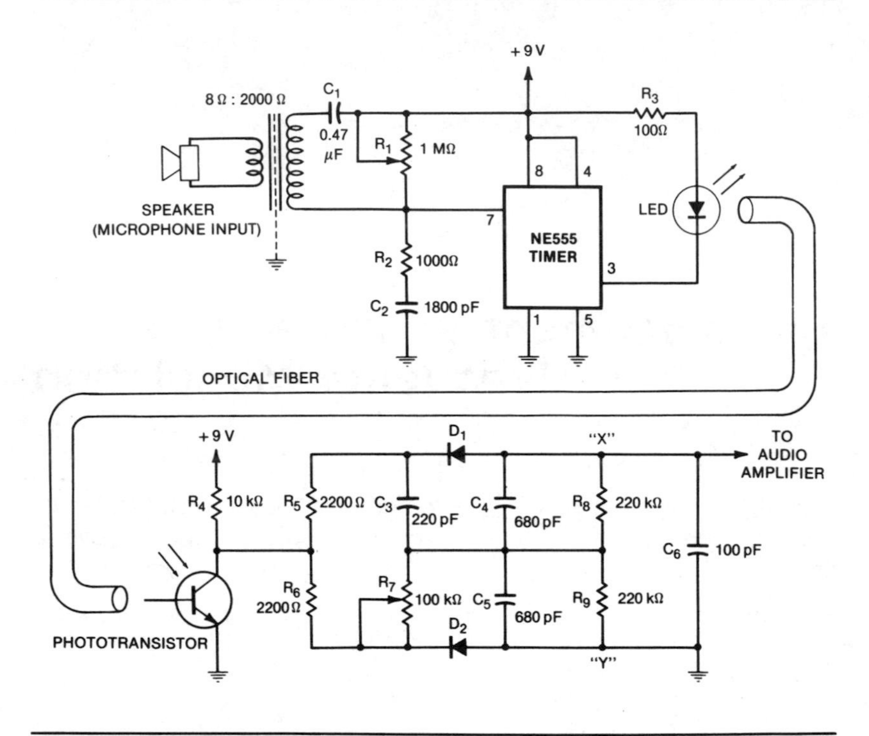

Fig. 14-1. Schematic diagram of pfm transmitter and receiver.

Materials Required

1 Audio output transformer, Archer 273-1380 or equivalent
1 1800-pF ceramic capacitor
1 220-pF ceramic capacitor
2 680-pF ceramic capacitors
1 100-pF ceramic capacitor
1 0.47-μF capacitor (any type)
2 2.2-kΩ resistors
1 100-Ω resistor
1 1-kΩ resistor
1 10-kΩ resistor
2 220-kΩ resistors
1 1-MΩ potentiometer
1 100-kΩ potentiometer

2 Signal diodes, Archer 276-1123 or equivalent
1 Phototransistor, Archer 276-145 or equivalent
1 LED, Archer 276-143 or equivalent
1 555 timer chip, Archer 276-1743 or equivalent
1 8-ohm dynamic speaker, miniature size
1 Fiber-optic thread or cable
2 Experimental boards or perforated circuit boards
1 Battery, 9-volt

Procedure

Step 1

Connect the 555 chip and related parts on one experimental board in accordance with Fig. 14-1. Check the operation of the circuit by connecting the battery and listening to the effect on the am receiver/amplifier. Only the slightest sound will accompany voice or music modulation. Further tests can be postponed until a suitable detector section is completed.

Discussion

The receiver circuits of previous experiments will not respond to frequency modulation. Changes in frequency can be detected by a discriminator. Detectors in fm broadcast receivers operate at an intermediate frequency of about 10 MHz. Our interest lies from about 20 kHz to perhaps 50 kHz.

We can construct a simple discriminator detector by making use of the fact that the reactance of a capacitor decreases with increasing frequency, and vice versa. The balanced detector circuit of Fig. 14-1 will produce zero voltage between points "x" and "y" when the capacitive reactance of C_3 equals the resistance of R_7.

With increasing frequency, the voltage drop across R_7 remains fixed while the voltage drop across C_3 decreases; the output becomes unbalanced so that "x" will become less positive than "y." With decreasing frequency, the voltage drop across R_7 remains fixed while the voltage drop across C_3 increases, again unbalancing the output, making "x" more positive than "y." An amplifier connected between "x" and "y" will respond to the varying voltage, which reflects the frequency changes of the transmitted pulses. Capacitor C_6 bypasses the pulse frequency.

Step 2

Connect the components of the discriminator in accordance with Fig. 14-1. Use a second board. Check diode polarity carefully.

Step 3

Connect the output of the discriminator circuit to a suitable audio preamplifier, or if your choice is a convenient hi-fi music-center amplifier, to its high-impedance input. While you can use the receiver (audio amplifier) that you constructed earlier, you may not appreciate the high-fidelity response capability of the fiber-optics and digital systems. Do not use headphones until all units are balanced and the circuits are working satisfactorily, to avoid possible ear damage from feedback howls.

Step 4

Connect the fiber from the LED to the phototransistor, and speak into the microphone.

Discussion

Chances are, you will hear very faint, garbled sounds that may be a response to your voice. It would be pure luck if your circuits were balanced and ready for proper operation. A convenient median frequency (steady state, no modulation) must be chosen for the transmitter to match the characteristics of the balanced detector. The detector can be varied in balance over a limited frequency range with R_7, to match the median frequency chosen at the transmitter with R_1. Your experiment will not be successful until you have found the frequency that will produce zero signal at the detector output with no modulation.

Step 5

Place a small transistor radio, tuned to a station, in front of your microphone. Simultaneously vary R_1 and R_7 slowly, while listening for an improvement in response in the received signal. When a median frequency near the balance frequency of your detector is reached, a sudden increase in the receiver volume will occur. The detector response is quite sharply defined, and thus it may be a bit difficult to locate the best operating point.

Step 6

When you find a "working frequency," fine-tune potentiometers R_1 and R_7 for the best-sounding response from your amplifier. You will find that R_1 will have a very limited range over which to balance for optimum results.

Commentary

The frequency at which the prototype system worked best was found to be 51.6 kHz. Components vary in values from 5 to 10 percent. If you have a frequency counter or oscilloscope at your disposal, 51 kHz would be a good starting point, but your system might perform best 10 percent up or down from this frequency. Above 20 kHz, the exact pulse frequency is not important, and it should be selected to give the best response in the amplifier consistent with the least noise.

The prototype system worked well, its fidelity limited only by the input transformer and the quality of its components. If you wish to experiment further with this setup, you might obtain improvement in music fidelity by eliminating the input transformer and making your resistance changes across R_1 by other means. Don't overlook the potential effect of varying C_2 of the transmitter circuit at an audio rate.

In any case, once your system works, you have successfully transmitted a frequency-modulated pulse signal via fiber optics and demodulated it for interpretation at the receiver. Your fiber can be any convenient length consistent with the energy output of the LED and the sensitivity of the photodetector.

Section 3
Projects

PROJECT 1: AMPLITUDE-MODULATION FIBER-OPTICS RECEIVER
PROJECT 2: AMPLITUDE-MODULATED FIBER-OPTICS TRANSMITTER
PROJECT 3: FIBER-OPTICS LIGHT-TRANSMISSION CABLE
PROJECT 4: FIBER-OPTICS LIGHT-PEN CABLE
PROJECT 5: SINGLE-FIBER PASSIVE LIGHT PEN

15 Project 1: Amplitude-Modulation Fiber-Optics Receiver

In Experiment 2, we breadboarded an amplitude-modulation receiver that worked quite well. For the serious builder, and for a more permanent, long-range purpose, a completed eye-appealing, compact unit is much more satisfying and practical in many respects. Our project, as shown in Fig. 15-1, is neat and clean, with a performance that rivals a commercial unit costing much more.

Catalog items of this kind can be so tiny as to invite wonder that they perform so well. A typical receiver rests comfortably in a tablespoon; its fiber-cable connector and audio jack are almost as large as the receiver proper. Such miniaturization would be admirable and appreciated in a setting such as a telephone terminal, where many thousands are needed on an equipment rack. For ordinary use, however, ultramidgets are unnecessary.

Our receiver is fitted into a 3.25 in × 2.25 in × 1.125 in plastic box with a metal cover. The input jack containing a phototransistor is mounted at one end, and an audio jack and battery switch are installed at the other end. Between these is a perforated board upon which are wired about a dozen circuit components, including a thumb-wheel volume control for occasional special settings (Fig. 15-2). Two long-life 9-volt batteries are wedged between the circuit board and the switch, held in place by the metal cover. The receiver as illustrated has a total weight of about 5 ounces with both batteries installed in the case.

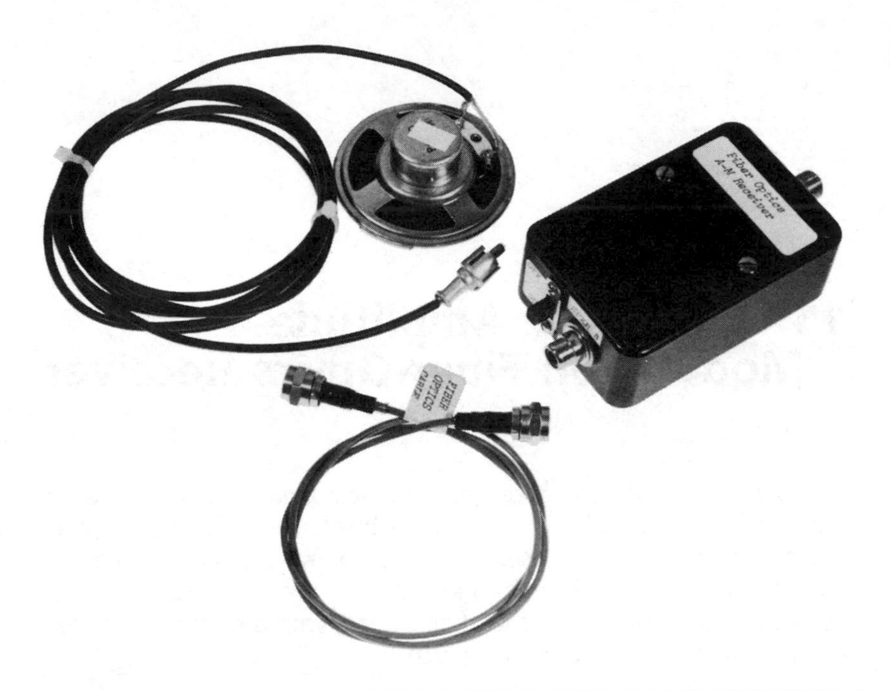

Fig. 15-1. Fiber-optics amplitude-modulation receiver.

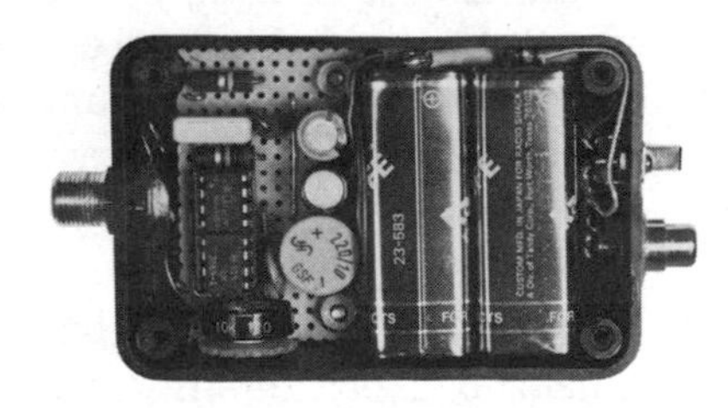

Fig. 15-2. Parts placement in fiber-optics am receiver.

Components

Receiver components are listed in Table 15-1. All are standard off-the-shelf electronic parts available from Radio Shack and stores featuring Calectro brand parts, among others. Variations will undoubtedly occur, but these will be superficial and easily worked around by anyone handy with electronic parts and tools. Take advantage of your "junk box" to the extent possible, obtaining the thumb-wheel potentiometer, for instance, from an old television chassis. Mail-order houses feature all of these parts, but many such outlets have minimum-order requirements. If you plan carefully for a number of projects, you may possibly benefit from prices as low as half as much as you would pay when buying them prepackaged from walk-in stores.

Long-life batteries are well worth their extra cost. Don't settle for a "cheapie" battery for this solid, professional-level project. Vero circuit-board products are slightly more expensive than the standard board listed in Table 15-1, but many builders of similar projects prefer them for their versatility; they are about as close to a printed circuit as you can get without actually designing and developing your own.

Preconstruction Considerations

You can put this fiber-optics receiver together in one or two productive sessions at your workbench. Perforated board was chosen rather than designing a printed circuit board for just one unit. Components are wired beneath and jumpered above the board/chassis. If you decide to build a number of these items, you can design your own printed circuit by following the pattern taken by the wiring, including the three or four jumper wires on top of the chassis.

By design, the volume control does not have an external means of adjustment. Once set at its operation point for your particular use, there is rarely any need to change its position. Should external adjustment be desirable, it is a simple matter to drill a screwdriver hole in the side of the case to provide access to the slot at the center of the thumb-wheel.

With the cover in place, the receiver is dust-proof, but not humidity-proof. Except when exposed to rain or spray from a garden hose, or to temperatures much above 50° Celsius, it is a solid performer, reliable and sturdy.

The receiver will respond to amplitude-modulated light in the wavelength range determined by the type of phototransistor you have

Table 15-1. Parts List for Fiber-Optics Receiver

Number on Diagram	Quantity	Description
R_1	1	220-kΩ, ¼-W resistor
R_2	1	1-kΩ resistor
R_3	1	10-kΩ resistor
R_4	1	10-kΩ thumb-wheel potentiometer, vertical mount, Archer 271-218 or equivalent
C_1	1	0.1-μF, 10-V capacitor
C_2	1	83-pF ceramic capacitor
C_3	1	220-μF, 10-V electrolytic capacitor
C_4, C_5	2	100-μF, 10-V electrolytic capacitor
IC_1	1	741 op amp, 8-pin DIP, Archer 276-007 or equivalent
IC_2	1	LM386, 8-pin DIP, Archer 276-1731 or ECG 823, or equivalent
S_1	1	Dpdt miniature slide switch, Archer 275-407 or equivalent
J_1	1	Tv coaxial bulkhead jack, F61A, Archer 278-212 or equivalent
J_2	1	Phono jack, panel mount, Archer 274-346 or equivalent
Q_1	1	Phototransistor, Archer 276-130 or equivalent
	1	Plastic box, 3.25 in × 2.25 in × 1.125 in, Calectro H4-723 or Archer 270-230, or equivalent
	1	Prepunched perforated board, holes 2.54 mm × 2.54 mm (0.1 in × 0.1 in), Archer 276-1395 or equivalent
	1	DIP socket, 16-pin, Archer 276-1998 or equivalent
	2	Batteries, 9-volt long-life or alkaline
	4	4-40 machine screws with nuts
		Hookup wire, stranded, small-diameter

selected. The prototype was built with an infrared-responsive component because the fiber-optics cable available and on hand was designed to operate best at infrared wavelengths. Unless the phototransistor is selected to match the cable characteristics, any attempts to use the receiver in a practical manner will be disappointing.

At this writing, fibers and fiber cables are not available off-the-shelf at your local walk-in electronic parts store. Some effort will be necessary to find suitable specific types by mail order. A number of known sources are listed in Appendix D, but you should study the advertising pages of the popular magazines and inquire by phone or letter if there is any indication that fibers and fiber-optics accessories might be available. A letter to the magazine editor may well turn up a few new sources not yet published, since there is at least a two-month time lag in preparing a magazine for publication. Phototransistors and LEDs are readily available from dozens of mail-order houses, as well as from walk-in stores.

Voice-modulated light is detected and amplified from about 10 Hz to 1 MHz with the prototype unit. Thus, you might receive high-fidelity music from a distance of up to a kilometer (0.8 mile) or more. The unit would not, in all probability, be adequate for reception of color television signals, except for the audio portion of the transmission. It would perform quite well in ham-radio slow-scan television circuits, in teletypewriter work, and in some of the slower facsimile machines.

Battery drain during operation will depend upon the amplitude at which the speaker (if used) is operated. As a preamplifier for a powerful hi-fi amplifier, for instance, the drain for the audio section is about 2 mA; for the phototransistor and first operational amplifier (op amp), the drain is only a few microamperes. Thus, battery life will be quite long, particularly for the battery supplying the negative voltage to the 741 op amp; the useful life of this battery should approach the shelf-life.

The receiver is straightforward and simple in circuitry and construction. Light enters the transparent window in the phototransistor mounted in the cable-connector jack, causing the npn "chip" to have increased current conduction in an approximately straight-line relationship with the increase in light level. A voltage that reflects the current increase is developed across R_1 (Fig. 15-3), and this voltage is in turn connected to the input pin of IC_1 through C_1.

The op-amp output is applied across R_4 to pin 3 of IC_2, an audio amplifier "chip." The level of the audio signal available from pin 5 of IC_2 is sufficient to drive a small speaker (up to ¼ W). To conserve battery power, you may wish to use a follow-on power amplifier instead of a speaker. Use an amplifier with a low-impedance input, from 8 ohms to a high of about 300 ohms, to match the low impedance of the receiver output.

Construction Notes

Batteries are included within the receiver case to provide a compact, totally self-contained light receiver. If desired, a suitable three-circuit jack and plug might be incorporated at the audio-output end of the case, for connection to a dual-voltage power supply to replace the batteries. This would make possible a smaller package.

Construction requires only hand tools common to the experimenter's work table. Care should be exercised at every step, particularly be-

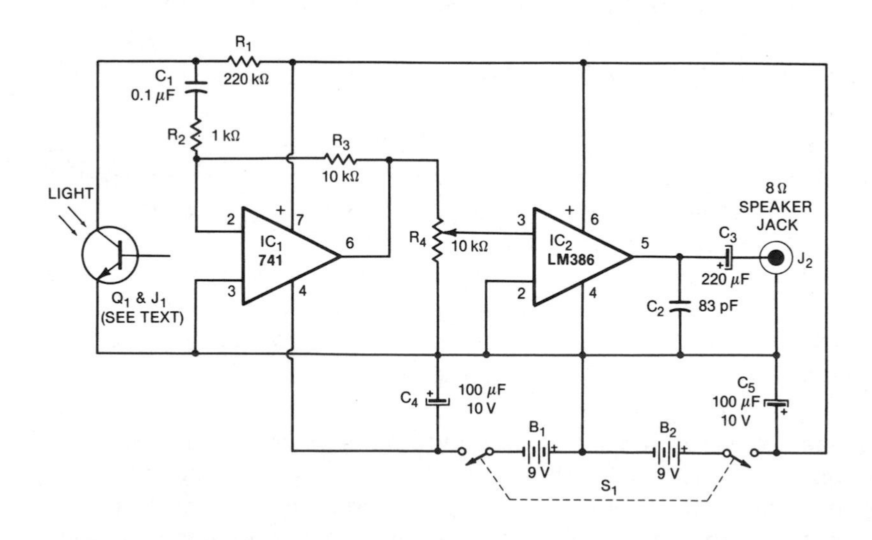

Fig. 15-3. Schematic diagram of the fiber-optics receiver.

cause the case is made of very soft, low-temperature thermoplastic. A small metal box will serve nicely instead of the plastic case, but a metal container may not have the neat external appearance the black plastic provides.

Fiber-optics cable jacks and plugs are somewhat expensive and usually are hard to find except in large quantities. A quite satisfactory substitute for a manufactured item can be made from a standard tv coaxial-cable jack and plug shell, as noted in Step 2 and in the cable project following. If professional-quality fiber-optic connectors are desired, they are available from AMP and Motorola (see Appendix D), but the cost for the connectors and crimping tools may put these items beyond all but the professional budget.

Construction Procedure

Step 1: Preparing the Case

First, you will need to drill the necessary holes in the case for the fiber-optic connector, the speaker connector, and the switch. In the prototype, a plastic case was used; your procedures may vary, depending on the size of your case and the material from which it is made. It is preferable to start with an index or pilot drill that is smaller than the desired hole. Larger drills are used successively until the hole is just about the right diameter. A small round file or long-taper reamer may be used to bring the hole to the required diameter.
The rectangular switch hole can be made by drilling two or three small holes in a vertical pattern, and cutting out the material between them with a small, sharp knife. The hole may be finished by using a small square or flat file. The rectangular hole may also be made by drilling a central hole and filing the hole to the required size.

The switch is mounted with the flange outside the box to provide extra room for the batteries inside the small case. The switch terminals are clipped off very short to permit the wires to fit between the inside wall of the case and the batteries. If the switch mounting flange is threaded for machine screws, the threads can be removed by drilling out the mounting holes with a drill just large enough to strip out the threads without appreciably increasing the size of the holes. Nuts and bolts will be used to mount the switch in the case.

Step 2: Modifying the Cable Jack

The cable jack is modified from a standard tv coaxial-cable jack. The insulated female pin is removed along with all filler by cutting off the crimped part of the shorter mounting portion of the jack and pressing the insulation out of the jack. A small lathe is perfect for cutting the crimped portion, but a satisfactory job may be done by drilling into the end with a 5/16-inch drill just enough to cut away the crimping. The plastic insulation may then be pushed from the long end toward the shorter end with a suitable rod (use something with a flat end, such as the reverse end of a drill bit, to avoid producing a wedging action on the soft material). If this fails, drill through the center axis with a 3/16-inch drill, and then dig the material out.

With the mounting shell empty, insert the phototransistor into the hole, and estimate the amount of build-up required to form a snug concentric fit. Select a phototransistor packaged with the sensitive material and window on the end rather than on the side.

Wrap a few layers of narrow masking tape or tape such as Scotch Magic Mender tape around the transistor, keeping the front edge of the tape even with the top edge of the phototransistor. When you have enough wound on the body to fill the mounting shell, insert the transistor, window forward, until it rests firmly against the forward stop rim. Do not use ordinary transparent tape nor electrical plastic tape, as layers of these tend to creep or shift over a period of time. You may obtain excellent tape for the purpose in very narrow widths from the drafting section of any large stationery store.

Set the shell on a smooth, flat surface, with the transistor face down. Fill the recess in the back of the jack with epoxy or other dielectric sealant, leaving the transistor leads extending straight up for the time being. When you have finished this operation, set the assembly aside to cure for at least two hours.

Step 3: Preparing the Board

Cut a section of perforated board that will fit into your enclosure or chassis without difficulty. If you are using a small enclosure such as the one shown in Figs. 15-1 and 15-2, be sure to leave enough room for the batteries, jacks, and power switch. Perforated board material can be cut by scoring it with a sharp knife and then breaking the board by aligning the scored line along a sharp edge (a counter top or table will work well) and pressing on the edges of the board.

Drill any required mounting holes so that the board can be mounted securely to the enclosure or box. Standard 4-40 machine screws are small enough to be used. A spacer is also recommended so that conductors on the back of the board do not come into contact with a metal chassis, and so that they are not bent, causing short circuits among themselves.

Step 4: Mounting the Components

A typical arrangement of parts is shown in Fig. 15-2. Your parts placement will probably vary to suit your needs. The main point is that the components should be neatly positioned and not crowded against each other. A crowded arrangement of parts will make troubleshooting, if necessary, quite difficult. The parts arrangement is not critical to the operation of the circuit, although you may wish to position the thumb-wheel volume control close to one side of the box.

For convenience and saving in cost, a 16-pin dual–in-line–package (DIP) socket is used rather than two 8-pin sockets. For wiring and soldering, the numbering will be 1, 2, 3, 4, 1, 2, 3, 4 rather than 1, 2, 3, 4, 5, 6, 7, 8 along the first side (Fig. 15-4). *Caution:* Don't forget; the numbering is 8, 7, 6, 5, 8, 7, 6, 5 along the other side from the same direction, corresponding to the pin numbering for the integrated circuits and not to the pin nomenclature in most diagrams of these sockets!

Insert the DIP socket, and bend two end pins down to keep it from falling out. If desired, a bit of glue on the bottom of the socket could be used instead to hold it in place, leaving the pins all straight after they exit through the holes. Insert the thumb-wheel potentiometer, and bend all three pins down against the board.

Be sure to observe the polarity on each of the electrolytic capacitors. These may be held in place with a spot of epoxy cement. Use the

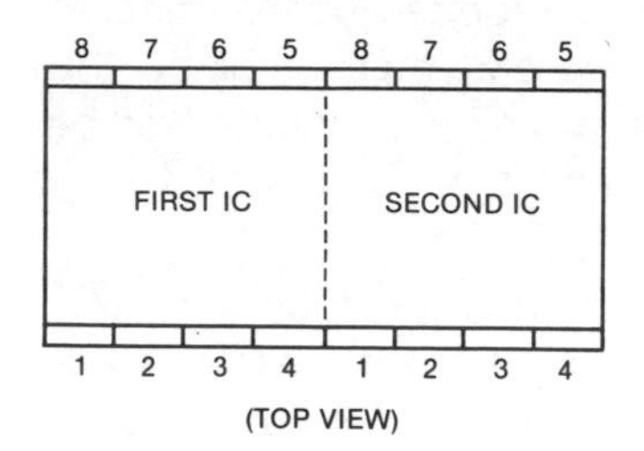

Fig. 15-4. Socket pin numbering sequence for two 8-pin ICs in one 16-pin socket.

cement sparingly so that it does not make a mess of your board. The epoxy should cure in about 30 minutes.

Once the epoxy cement has cured, you can begin to wire the components in accordance with Fig. 15-3. Use short pieces of jumper wire to make the connections, and solder them securely. You may wish to make up two short wires with one-turn coils or eyelets on one end. These can be positioned vertically and soldered in place at the phototransistor end of the circuit board so that the phototransistor can be easily connected once it and the board have been fastened in place. Since commercially available pins for perforated boards are fairly large, they were not used in constructing the version of the receiver illustrated here.

Recheck your wiring to be sure that it is correct. When you have inserted the integrated circuits, be sure that the pin orientation is correct before going further.

Step 5: Installing the Phototransistor and Jack

Before inserting the transistor-jack assembly into the case, check the leads with an ohmmeter to determine which lead is to be connected to the positive battery supply. The meter should show a resistance change with the application of light to the transistor window. On most multimeters, the positive (red) lead indicates the positive lead on the phototransistor when the resistance changes between light and dark conditions are measured. If there is no indication of change between light and dark, switch the test leads.

Some types of digital ohmmeters will not read in this manner, as they do not place enough voltage across the unit being tested to cause it to conduct. In this case, use a 1.5-volt dry cell in series with a 1000-ohm resistor, and measure the voltage across the phototransistor on the low voltage range of your multimeter. The ''positive'' side of the phototransistor will be the one that is connected to the dry-cell positive terminal when the phototransistor reacts to light.

Mark the positive lead with a touch of ink or paint, or twist a tiny loop into it. This lead will be soldered to the terminal nearest the input capacitor, C_1 in Fig. 15-3.

Place the lockwasher on the jack, and insert the jack into the case. Now, place the nut on the threads and tighten. The lockwasher is used to help prevent the jack from loosening when the cable assembly is removed and replaced from time to time, so be sure the lockwasher is on the *outside* of the case. Orient the jack to place the transistor leads in a horizontal plane with the box lying flat on the bench. The positive, marked, lead should lie on the side toward the input capacitor. Bend the leads so that the ends just enter the eyelets curled into the vertically standing terminals, but do not solder them at this time.

Step 6: Installing the Audio Jack and Battery Switch

The audio jack is installed in the case. In the system shown, the jack was too long to fit in the case with the other components, so some of the contact for the inner conductor was removed. Both this contact and the ground tab were tinned with a small amount of solder to make soldering the wires to the jack a bit easier.

The battery switch is secured in place with machine screws. Be sure to mount the switch flange from the outside of the case if you are using a small enclosure. This will give you additional room inside the box.

Step 7: Final Wiring

Once the components have been mounted and wired on the board, and once the chassis-mounted components have been installed, you may proceed with the final wiring of the fiber-optic receiver. Use of small-diameter, stranded hookup wire is suggested, since you can place it as needed without fear of breaks in the wire and with a great deal of flexibility. Be sure to observe the correct polarity of the various signals and components as you complete your wiring. Use a minimum amount of heat on the battery, switch, and phototransistor terminals, since excess heat may damage the components or the connections to them.

Step 8: Final Wrap-Up

Since a hole for adjustment of the thumb-wheel potentiometer is a possibility, depending on your particular needs, you might wish to drill it at this time. Locating the drilling spot can be done with an outside caliper (a tool resembling a compass, but with legs bent "bow-legged"). Drill the hole, if you opt for one, before installing the board.

Install the circuit board. It will be fairly easy to do if you first insert the two 4-40 screws from the outside in and lay the box on a smooth, hard surface to hold the screws upright. Lower the two spacers over their respective screws; then work the board down over the screws. Follow with 4-40 nuts and tighten snugly. Give the two phototransistor leads a final curl to bring them into their terminal holes on the circuit board, and solder.

With the batteries lying conveniently outside the case, solder the four color-coded leads to their respective plus nine-volt and minus nine-volt terminals, the common terminal, and the audio jack.

Jockey the batteries into place. The wires that will pass across the battery from the switch may be placed between the upper and lower sets of battery terminals, which form a "channel" when the batteries are pressed into the box. Tuck the leads neatly behind the pylon.

An alternative to soldering the leads to the battery terminals directly is to dismantle two old batteries and remove the terminal caps by drilling the rivets that hold them to the insulating strip. The caps may be soldered to the leads, providing clip-on connections to the battery terminals. Space across the box is at a premium. The Calectro box (or other slightly larger box) is a shade larger and will accommodate the clips to provide removable caps for battery terminals. If you have an appropriate cutting tool for the job, you might shave a thin sliver from the inside of the Archer box to give a bit more room for the contacts.

If you have made all connections very carefully, your fiber-optics receiver is now ready for its first test.

Step 9: Testing

Connect a small speaker to a phono plug and cable, and plug into the audio jack. Turn the switch on.

Point the receiver phototransistor toward a fluorescent light, or toward the lighted LED of a hand-held calculator (liquid-crystal readouts will not produce a signal).

Turn the volume control until a hum is heard. The volume is surprising if your receiver is functioning well.

If no sound is heard, recheck all connections from Step 1, and check each component for correct polarity, solder connections, and location in the circuit.

Did you forget to press the switch on? Did the switch slide make contact? (If the switch mounting nuts are too large, they can interfere with switch slide action.) Does −9 volts appear at pin 4 of the op amp?

Step 10: Final Assembly

Using the screws provided with your case or chassis, install the cover. Turn the unit off until you decide to use it in an application. This completes the project.

16 Project 2: Amplitude-Modulated Fiber-Optics Transmitter

The fiber-optics transmitter is almost exactly the same in appearance as the receiver constructed as the first project. Externally, the same type of box is used, as shown in Fig. 16-1. Note that the arrangement of the fiber-optic jack, audio jack, and power switch is almost the same as that found in the receiver unit. Press-on labels are suggested so that you can easily tell these units apart.

Inside, the projects are very similar, too, with a similar arrangement of the major components. A similar-size piece of perforated board has been used for the construction of the transmitter circuitry.

The circuit used in this project is simply an audio amplifier feeding a transistor, which in turn modulates the light output of a light-emitting diode. The LED is contained in a modified coaxial-cable connector shell.

Components

The components used in this project are listed in Table 16-1. The parts are readily available, and you can probably find many of them in your "junk box" of components salvaged from other projects and circuits. The choice of an enclosure is up to you, depending on your needs. We used a small plastic case to "miniaturize" the transmitter. An infrared LED has been used to match the receiving phototransistor used in the previous project.

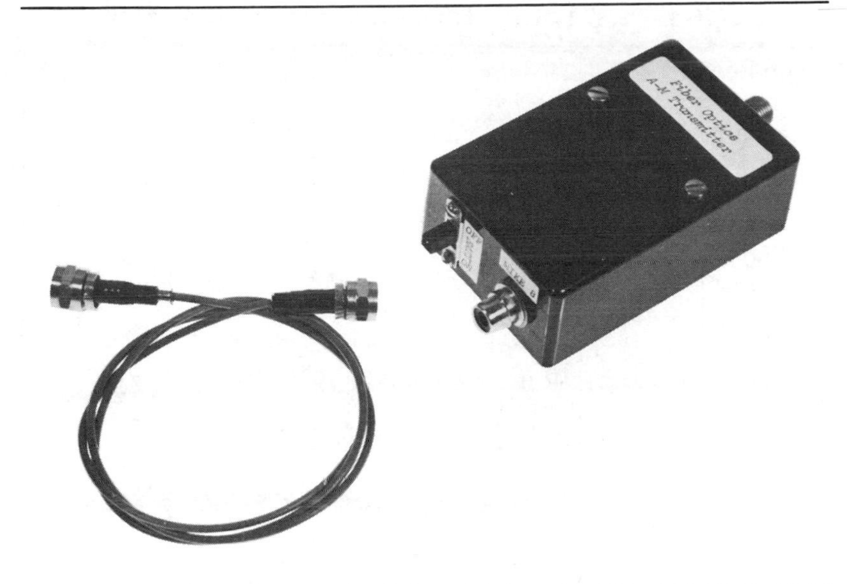

Fig. 16-1. The fiber-optics transmitter module and fiber.

Construction

The construction of this project is similar to the construction of the fiber-optics receiver described in Project 1. Therefore, we will not provide all of the mechanical details for you here, referring you instead to the first project for the general information. If you have constructed that project, you are already aware of the techniques and materials used. It is recommended that you build the receiver first. The circuit for the transmitter is somewhat simpler than the receiver circuit, so you will find that your circuit board is a bit less crowded.

An inside view of the transmitter is shown in Fig. 16-2, and the schematic diagram of the circuit is shown in Fig. 16-3. Rather than mount the transistor through holes in the board, unused contact pins on the 14-pin DIP socket have been used. This allows other transistors to be substituted, if you wish to do so.

Step 1

The first step is to prepare the chassis or box by drilling and sizing the three holes needed for the fiber-optic jack, the audio jack, and the

Table 16-1. Parts List for Fiber-Optics Transmitter Module

Number on Diagram	Quantity	Description
R_1	1	1-kΩ resistor
R_2	1	50-kΩ resistor
R_3	1	1-MΩ miniature thumb-wheel potentiometer, Archer 278-212 or equivalent
R_4, R_5	2	100-ohm resistors
C_1	1	0.22-μF low-voltage capacitor
C_2	1	10-μF, 10-V electrolytic capacitor, vertical-mount pc-board type
IC_1	1	741 op amp, Archer 276-007 or equivalent
Q_1	1	Transistor, npn, Archer 276-2009 or equivalent
LED	1	Infrared or visible type, Archer 276-142, Archer 276-143, or equivalent
S_1	1	Dpdt switch, miniature slide type, Archer 275-407 or equivalent
	1	Phono jack, panel mount, Archer 274-346 or equivalent
	1	DIP socket, 14-pin, Archer 276-1999 or equivalent
	1	Tv coaxial jack, chassis mount, F61A, Archer 278-212 or equivalent
	1	Prepunched circuit board; Archer, Vectorboard, Veroboard, or equivalent
		Miniature-size stranded shielded hookup wire
	4	4-40 machine screws with nuts
	2	Spacers, about ¼ inch
	1	Box, black plastic, 3.25 in × 2.25 in × 1.125 in Calectro H4-723, Archer 270-230, or equivalent
B_1, B_2	2	Batteries, 9-volt, long-life or alkaline

switch. Once these holes have been made, you may need to "modify" the switch and the audio jack as noted in Project 1, so that they will allow sufficient room for the batteries and circuit board.

Fig. 16-2. Inside view of fiber-optics transmitter.

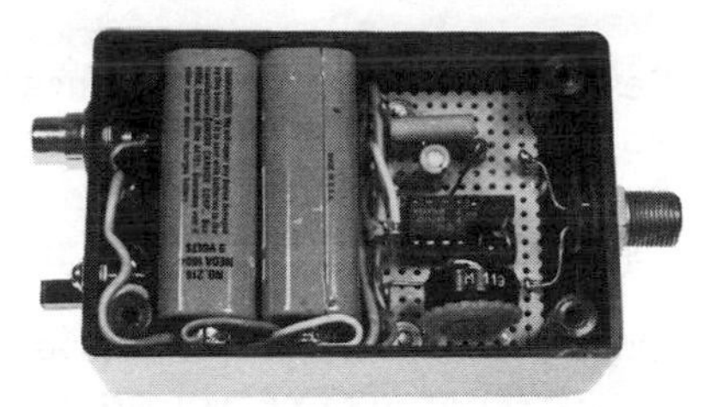

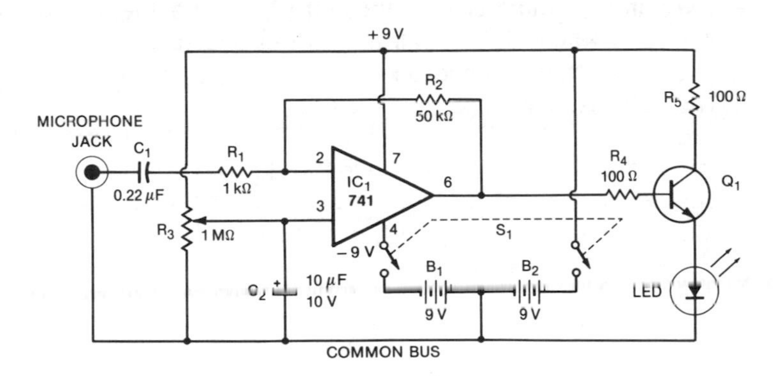

Fig. 16-3. Schematic diagram of fiber-optics transmitter.

Step 2

After preparing the case, you must modify the coaxial cable jack so that the LED will fit into it, and so that the LED can be permanently mounted. The modification technique has been described. Of course, an end-emitting LED must be used. You must build up the diameter of the LED with tape so that it fits snugly and concentrically into the modified jack.

The entire microphone input lead should be shielded; the prototype had severe feedback until this was done. Use miniature shielded wire because the larger variety takes up too much room. The remaining construction details are left to you. They closely parallel those provided for Project 1. You will probably modify your enclosure and circuit board from that shown, so further specific discussion is not required.

Operation

Before you attempt to use your transmitter, review all of the connections on the schematic diagram so that you are sure that they have been made correctly. It is often easy to overlook the polarity of batteries, diodes, transistors, integrated circuits, and other components.

Once you are sure that you have all of the connections properly made and soldered in place, you are ready to test the fiber-optic transmitter. Before you proceed to actual testing, however, be sure to note the following:

A. Capacitor C_2 must charge through R_3 before the system becomes operational. This will take a few seconds from the time the transmitter is turned on.

B. The LED may have been installed with its leads reversed in the circuit. The LED will not function if the leads are reversed, but this will not harm it, either. Since the LED is probably an infrared emitter, you cannot tell if it is operating properly by looking for emitted light. If you suspect that reversed LED leads are causing a malfunction, simply unsolder them and resolder them in the reversed positions. The jack may be rotated, if necessary, to account for repositioned LED leads.

When testing the transmitter (and the receiver), be sure to take your time. Work patiently and carefully during the following steps.

Couple the transmitter and receiver modules by placing the two fiber-optic jacks so that they are pointing at each other, end to end, and so that they are as close together as possible. Wrap the jacks with black electrical tape to make a fairly good mechanical connection and to exclude external light. This connection is crude, but it is sufficient for testing purposes.

We have assumed that the LED and phototransistor have their active elements or surfaces very close to the center line of the jack so that they are closely aligned. If you suspect that one or both may be off-center a bit, the receiver and transmitter modules may be rotated with respect to each other to get the best alignment.

Step 1

Plug in the speaker at the receiver audio jack.

Step 2

Plug in a microphone or another 8-ohm speaker at the transmitter audio jack.

Step 3

Turn the receiver on, and listen for a slight hissing sound, indicating that the receiver is operating.

Step 4

Press the transmitter switch on. Wait for at least 15 seconds, while listening for any change in sound in the receiver that might be caused by the transmitter. Rub your fingers across the microphone diaphragm (or rub a hard object across the microphone external protective screen), and listen carefully for receiver response. You are not likely to hear any great sounds at this time, so listen carefully in quiet surroundings. There is danger of the system breaking into very loud oscillation, with potential damage to your ears, so headphones are not recommended at this time.

Step 5

The adjustment of R_3 is critical. Start with the potentiometer in its minimum position. This will depend on your configuration and how you mounted the potentiometer. Rotate the potentiometer slightly, listening for a change in the output of the receiver. Be sure to wait for the system to respond to the adjustment, generally 15 to 30 seconds. Touching the microphone or the microphone-speaker diaphragm is a good way to "generate" test noise.

Step 6

If no sound is heard at the receiver by the time you have gone about three-quarters of the way around with the thumb-wheel potentiometer, return it to its starting position, reverse the LED leads, and repeat the last two steps. When the system starts to function, the resulting feedback "howl" will let you know in no uncertain terms that the system is working.

There is an optimum position for the thumb-wheel control. Find this position by isolating the audio output. One way to do this is to connect the microphone jack to the output from a low-level phono pickup playing a record.

If you have an audio function generator and an oscilloscope (best of all test instruments), use these to determine the best setting for the potentiometer.

Step 7

When the system is working to your satisfaction, solder the emitter LED leads if they are not already soldered in place. The LED fiber-optic jack may be tightened if you rotated it to reorient the LED leads. The chassis or box may be bolted together or closed. You may wish to label the transmitter and receiver, and the switches and jacks.

Troubleshooting

In case the system does not work, check all connections, and test all voltage points with your multimeter. If you are using an infrared LED, you cannot see when it is active; therefore other tests are required. An oscilloscope is excellent for such troubleshooting, but ingenuity with a multimeter can accomplish the same ends.

Check the battery-switch operation. If the switch has not been mounted correctly, the slide may not be making contact.

Sometimes a battery is faulty right from its package, because the positive terminal is often only a press connection to the cell on top. If you soldered your battery connections, you may have used too much heat and warped the insulating strip at the terminals.

Measure the battery voltages from the "common" terminal on the chassis to each side of the switch in the off position. Then measure with the switch on.

Be particularly careful in evaluating the connections to the 14-pin DIP socket. Use caution when inserting the transistor and IC into this socket. Remember that the terminal count beneath the chassis is the mirror image of what it appears to be on top, and the triangular placement of the transistor leads in the socket can be all too easily misinterpreted when viewed from beneath.

Once the transmitter is functioning and the receiver is likewise performing well, you can go to Project 3 with expectations that you will have an excellent fiber-optics system with hundreds of uses.

17 Project 3: Fiber-Optics Light-Transmission Cable

There are several ways to make a usable light-tight communications path between your transmitter and receiver. The two units can be connected permanently by cementing a fiber to the LED and phototransistor. The fiber or fiber bundle can be jury-rigged into contact with the light elements. Best of all is a carefully constructed cable with convenient removable couplings.

The critical need for concentricity among the light elements must be kept in mind. If the cable does not accurately match the "hot spot" in the phototransistor or LED, results will be disappointing at best. A single-strand cable is the most critical, requiring the greatest care in construction. Because there are four possible sources of off-center error, any offset condition may be multiplied many times.

It is a good idea to test each step of the following construction details with your live receiver before finalizing the step. Use the light output from a small LED-type calculator as your test transmitter, or you can use a pinhole in a piece of black paper placed over a fluorescent lamp.

Construction Notes

Table 17-1 contains a general list of materials for this project. Because fibers and fiber bundles are not yet supplied "off the shelf" with specific reference numbers, it is not possible at this time to direct you to buy a particular cable, except to say that the wavelength characteristics of the cable must closely match those of the LED and phototransistor. Instructions for assembly of couplings onto whatever cable you may acquire must therefore be somewhat general in nature.

In your search for filler material to use between the outside of your cable and the inside of the coupling, don't overlook the outer sheath of RG58U and other coaxial tv cables. Perhaps a metal or fiber spacer ordinarily associated with screw hardware will be just the right diameter and thickness. Wrapping with paper or other suitable tape to the desired diameter is workable, but it can introduce off-center conditions if not carefully done.

You can dress your final assembly somewhat by slipping a piece of shrink-tubing over the outer portion of the coupling, and shrinking it over the coupling and about an inch of the fiber-optics cable. Don't forget to slip such tubing down and out of the way over the cable prior to working on the final coupling!

Table 17–1. Materials List for Project 3

Quantity	Description
1	Small bundle of fiber-otpics strands (fibers). Single-strand fiber may be used, the larger the diameter the better. A cable such as the Galileo 1000 200-T cable has a cross-sectional diameter of 2 mm including outer sheathing and contains multiple strands of sufficient number to minimize concentricity problems.
2	Television coaxial-cable female couplings, Archer 278-214 or equivalent. There are two common shank sizes. Select the size that most nearly fits your cable.
	Tape, fiber spacers, or other filler materials.
	Miscellaneous selected sizes of heat-shrink tubing

When you have completed your placement of the coupling on the cable, but before you make it permanent with the use of cement, screw the coupling into the receiver jack finger-tight. Back off one turn, and press the cable firmly into the coupling. This will cause the cable to move into the coupling ferrule just a bit further. Then, when the assembly is cemented, subsequent connections to the receiver jack will bring the polished end of the cable up tightly against the window of the phototransistor. As you may be aware, there will be slight differences in positions between the light sensor of your receiver and the light emitter of your transmitter. Therefore, when you make the adjustment just described, it would be prudent to mark your cable with a distinctive color so that this end of the cable can always be connected to the jack for which it is adjusted. When you place the coupling on the other end of the cable, use the LED jack on the transmitter unit to adjust the spacing before cementing. Then, use the color coding so that you can always match cable and jack.

General Suggestions

Polish the fiber ends according to the instructions in Appendix F. This is vital because light will not enter or leave a dull, unfinished strand or cable end with sufficient power to give satisfactory operation of your system.

If you settle for a single strand of light fiber, insert it into the insulation from a piece of No. 18 or 20 bell wire or single-strand hookup wire. Many types of insulation are thermoplastic; that is, they soften when heated. This makes skinning insulation from the wire relatively easy. Place the bare wire end in a vise, and with a glove- or cloth-protected hand, stroke the full length of the wire with moderate squeezing pressure. Let the insulation slip through your fingers until it heats enough from the friction to soften and expand, whereupon it will begin to move along the wire. Continue full-length strokes until the insulation is fully removed. Insulation has been removed in this way from a piece of No. 18 bell wire 30 meters long. Of course, the wire should be larger than the fiber(s) that will replace it.

Inserting the fiber into a long length of such material can be a tricky task, but several methods can be employed innovatively to ease the difficulties. One is to introduce a fine sewing thread as a pilot by placing a vacuum pump at one end of the insulation and feeding the thread into the other end. When the thread comes out the vacuum end, it can be "instant cemented" to the fiber and the latter pulled carefully into the insulation.

If seven or more strands of fiber are bundled to construct a cable, the wire size from which you obtain the insulation will have to be larger. Do not overlook common single-strand electrical house-wiring (No. 14 or larger) as a source for outer sheathing.

Construction Details

Step 1

Using paper or other selected build-up materials, increase the size of your light cable until it fits snugly into the coupling. When you are ready to secure the cable to the coupling, instant cement will do the job nicely. (Note the foregoing general construction suggestion about how to make a good pressure fit of coupling to jack.) Refer back to Fig. 9-1 in Chapter 9 for refresher suggestions.

Step 2

Select a suitable length of shrink tubing, and slip it over the ferrule of the coupling; it should extend for a short distance along the length of the cable. Heat it for a final eye-pleasing touch. It will have the added utility of reducing the amount of sharp bending that could take place between cable and coupling. Sharp bends can all too easily snap the fiber in two—remember, it is made of glass.

Step 3

Place a similar piece of heat-shrink tubing over the other (open) end of the cable, and slip it down out of the way where it will not be forgotten. Prepare the open end of the cable for its coupling, and install in the same manner as directed above.

Wrap-Up

The cable is ready for use when you have completed Step 3 and the cement has had ample time to catalyze—provided the fiber ends match the light elements on receiver and transmitter, that is.

Couple the receiver and transmitter units together with the cable, and try the system. If you are satisfied with its operation, let well enough alone. If not, you may have to "tinker" with the cable, the couplings, and perhaps even the mounting of the light elements in their jacks. Be sure that the light fiber butts snugly against the light elements in the jacks of both transmitter and receiver when the screw nuts are tightened. This can be determined by backing off a half turn on the coupling and moving the cable slightly with your fingers as you slowly retighten. If your construction has been correctly done, you can sense the contact of cable to light element just before the coupling is finger-tight.

Do not forget that fibers are manufactured for different wavelengths of light transmission, from infrared to ultraviolet and beyond in each direction. If your cable does not match the characteristics of the phototransistor and LED, the system will work poorly, if at all, even though the couplings have been perfectly matched to the jacks. If, in addition to the cable, the LED and phototransistor are of unknown make and characteristics, you have considerable empirical matching to do, and this should be done before the close work of constructing jacks and couplings is undertaken. It may be wise, therefore, to buy a selection of each from a catalog listing, rather than a specific item, because cable is not generally available to walk-in purchasers as yet. When the match is made and the couplings are concentric, the system will perform with surprising fidelity.

18 Project 4: Fiber-Optics Light-Pen Cable

There are occasions when a means of inputting light into your receiver from unusual sources is desired. A "jury-rig" is possible in most instances, but the convenience of a cable with a coupling on one end is both a time saver and an assurance that you are getting maximum transfer of light from source to receiver. It is even more important to have a good, light-tight coupling when the source is barely detectable.

While a cable with a coupling on one end and only the polished end of a fiber on the other would serve as a usable test lead for light transfer, utility and aesthetics are both served better by enclosing the probe end of the cable in a protective sheath. A felt or fiber-tipped pen makes an excellent test prod for your cable (Fig. 18-1), and it requires only a minimum of rework to adapt it to its new role.

A "light pen" of this kind makes your fiber-optics am receiver a handy test instrument in its own right, and the receiver also may be used to input various signals into other test equipment. For instance, you may wish to determine the speed of a motor shaft. With a light shining on the rotating shaft, the light pen will pick up the variations in light intensity caused by the different reflectivities that are inevitable in most machinery. The effect can be greatly enhanced by painting a white or black stripe longitudinally on the shaft, as appropriate. At the receiver, the sound emitted from the speaker can be compared with the output of an audio signal generator, and the revolutions per

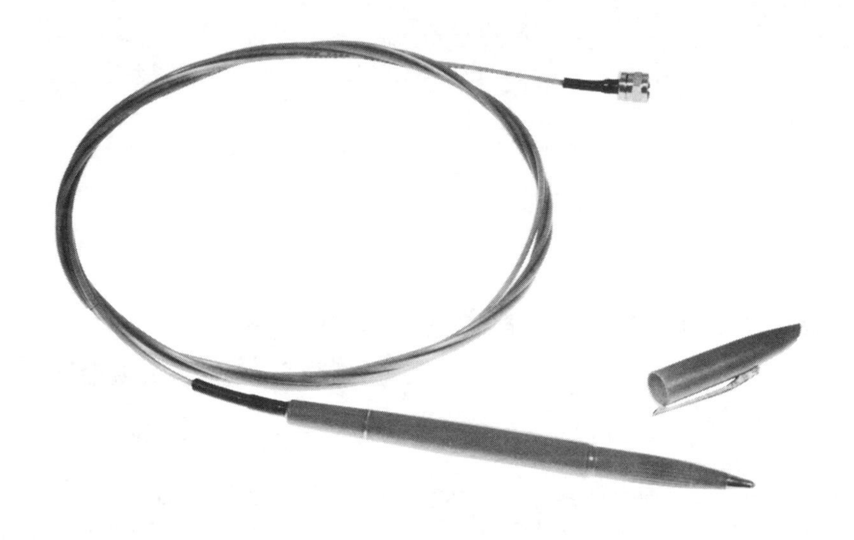

Fig. 18-1. Fiber-optic light pen with adapter for receiver jack.

minute can be determined. Another approach to such a measurement is to use the receiver output as the input to an event counter.

The buzzing of a housefly wing has been clocked on an event counter with the aid of this light pen. Ambient light was ample to register over 2000 beats per minute.

With the light pen placed at a specific spot on the screen of your microcomputer, you can cause your computer program to control numerous unconventional and remote activities. Your program can place a spot of light in any predetermined location and can make it blink or even follow or evade the light pen. The light emitted by the phosphor screen may seem to the eye to be steady-state, but in fact it consists of pulses occurring at over 15 kHz. Your light receiver will respond to this frequency either through a speaker or into other devices, such as a tuned filter and relay, which in turn can control the turning on or off of a motor or other electrical or electronic device.

Because thinking in terms of light as a communications medium is relatively new, we tend to overlook many interesting possibilities. But with the light-pen test cable handy, you will discover many applications that will suddenly occur to you at odd moments. That is the way our minds work: highly responsive to "what if's."

For this project, you have only to make another cable, as long as you wish and are able to find, and fit it with a coupling at one end as you did in Project 3. This project will begin with the cable in that half-completed condition and carry your second cable to completion as a handy "light-pen" test lead.

Construction Notes and Suggested Approaches

In a society that uses the telephone and face-to-face oral communication so consistently to the exclusion of letter writing, it is remarkable that so many different types of writing sticks are available in profusion. The ball-point pen has made the old cedar-wood pencil virtually obsolete, yet pencils may still be found in stationery stores. For this project, it is suggested that you visit a large stationery store and compare the wide variety of pen types with an eye to finding the one that best fits your needs.

A felt-tipped pen bearing the name "Sanford's Expresso" was chosen for the prototype, although the brand name is not important in this usage. Rather, you are seeking a pen that is easily converted into a test prod and that fits your idea of what a light pen should look like. Make a color selection to suit your preference. Examine the pen tips to find one that is not too large, and yet not so small that you will have to drill away the strengthening material to insert the cable.

Having selected the pen of your choice, open it and remove the "insides." If you have chosen a felt-tip pen, it will contain a cotton-filled reservoir of ink. Pull the cotton out, and wash the inside of the pen body with water and a bit of dishwashing detergent. If you select a pen with oil-based ink, other solvents such as alcohol or gasoline will have to be used. Check each item on a test surface of the pen before using, to make sure the liquid will not dissolve the pen body itself.

Pull the felt tip out from the inside, or press it with a blunt rod, and wash and dry the chamber thoroughly.

If you have chosen a ball-point pen, in some cases the ink carrier will respond to being pressed from the inside out, and in other cases the carrier must be pressed from the outside in to be removed. A more expensive pen with a disappearing tip usually comes apart by unscrewing the two halves of the pen body, and the replaceable reservoir and ball point lifts out easily. These pens would require little or no washing, since you would not be using the ink reservoir and ball-point tip.

Construction Details

Step 1

Prepare your light-fiber bundle for fitting into the pen body. Since innumerable configurations are possible, both of the cable and of the pen, specific instructions are likely to fit some needs and miss the mark entirely with others. In brief, the preparation should include slipping two or three pieces of flexible spaghetti or hookup-wire insulation over the fibers. Each piece should fit snugly over another in increasing diameters; the piece with the smallest diameter should be the longest.

Step 2

If appropriate, drill a hole in the pen at the end opposite the tip, to accommodate the largest piece of spaghetti. The point at which the fiber enters the pen body will usually be the top, but it can be along the upper side if desired.

Step 3

Check the diameter of the pen tip relative to the exposed fiber-cable end. Either drill out the tip or build up the fiber with tape, whichever is needed to make a snug fit in the hole. Adjust the amount of fiber extending through the tip to about ⅛ inch, and with the pen inverted, touch a drop of instant cement to the tip assembly. Make sure that the cement touches each layer of any tape buildup, including the point at which it emerges from the pen tip.

Step 4

When the cement has had about 20 minutes to cure, polish the fiber tip, using the procedure outlined in Appendix F. Your light pen is now ready for use.

Using the Light Pen

If your receiver contains an infrared phototransistor, your light pen will be used primarily for this wavelength, since it will be less sensitive to visible light. There are innumerable sources of infrared light in the environment, but most of them are constant-level sources to which the receiver will not respond without variation of the light entering the tip. For these, you might construct a chopper to couple between the cable and receiver, using, perhaps, a mirror and lens system, or the motor-driven shutter from an old 8-mm movie camera.

Fluorescent lamps emit some infrared radiation, but not nearly as much as does the common filament lamp. Filament lamps will excite the receiver because they are usually powered by the 60-Hz line, although the larger the wattage of the lamp the less variation in light output will be encountered. If you use the lamp as a light source and interpose an object between the source and pen tip, variations will be introduced as you sweep the tip across the object (opaque objects will not work here, of course).

If your receiver phototransistor is of the visible-light variety, there are hundreds of sources to be detected by your test probe. Again, however, the only response that a steady source will cause in the receiver occurs when the pen is swept across the source.

Don't overlook the computer screen as a source of active light input to your test lead.

19 Project 5: Single-Fiber Passive Light Pen

A passive light pen uses ambient light or heat to operate its amplifier and sensing apparatus. A single fiber carries a small quantity of light, and since it must sense that light from ambient sources, the brightness of the light is often at the very edge of the sensing ability of the photodevice. However, such a fiber can pinpoint a light source very closely. It can, for instance, discriminate between the lines and spaces on a small television screen. With the help of a small lens, it can discriminate between the dots in a color television picture. It can produce some interesting results when placed over the groove in a phonograph record with a strong light directed downward onto the record. An example of a single-fiber light pen is shown in Fig. 19-1.

A single fiber with limited light input would benefit from using a Darlington-type phototransistor in the receiver, or, if your budget permits, a more expensive type that employs secondary emission may be used. However, the receiver you have built will give some surprising responses. In any case, because of limitations beyond the scope of this book, the pen will not be a universal test instrument. Should you decide to extend the range and capabilities of your receiver, this project will perform very well as a test probe.

Constructing a light pen from a single fiber of the size suggested in this project is an exacting task. It is not to be entered into lightly, nor rushed into headlong as you might in constructing a multistrand light-fiber project. Your tools should be equal to the demands of the project, and not the least of these is patience. You may find a drill bit the size of a human hair rather difficult to handle. However, ingenuity can substitute for expensive tools in this work as in any other. A steel guitar string can, in a pinch, drill as neat a hole in brass as a well ground watchmaker's drill bit, and steel music strings come in quite a variety of sizes.

The pen can be used for a number of interesting purposes. By using it in conjunction with a counter, for instance, you can determine the speed of the pulses on calculator LEDs. You can "play back" the sound track on a strip of movie film directly into an amplifier or pick up the flashes of light from the teeth of a wrist-watch gear. With the appropriate accessory, you can stop a scanning sequence on a string of LEDs or play a game on your tv screen. You might even try to pick up the sound from a phonograph record, and, if your curiosity is piqued beyond resistance, construct a circuit that will separate the left channel of a stereo recording from the right channel (vertical vs horizontal cut). If the energizing light is supplied from a laser diode, your experiment can be intriguing indeed.

Materials Required

Corning Optical Waveguide, 2.9 dB/km at 850 nm, 39 MHz, No. 1505 (or equivalent)
Bic AF-49 Accountant's Fine Point ball-point pen (or equivalent)
Methyl ethyl ketone (MEK) *CAUTION*: This is a hazardous chemical.
Five-minute epoxy cement
Instant cement

Construction Details

Step 1

Remove the ball bearing from the pen tip. To do this, hold the pen vertically against an abrasive stone (such as Carborundum), and with a light, firm pressure make a number of circular sweeping motions. This will wear down the bearing, reducing its diameter so that it will fall out of its brass socket. A few flaps of the pen against a table may help dislodge a ball that remains stuck in the ink even after the diameter is

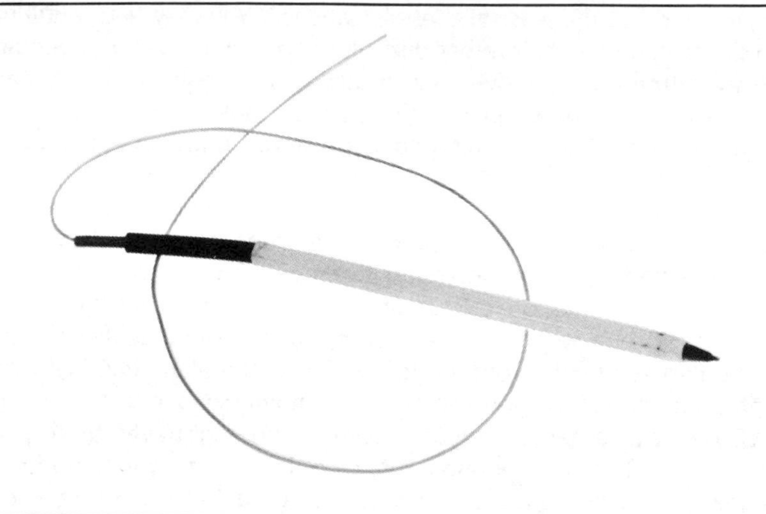

Fig. 19-1. A single-fiber light pen.

reduced enough to slip through the crimped socket. Be careful not to damage the brass tip.

Step 2

Clean the ink out of the tip and reservoir, using methyl ethyl ketone (MEK) or other solvent, obtainable from paint stores.

WARNING

Observe the precautions on the MEK container. Use in a well-ventilated area, and have no open flame within 5 meters (15 feet). Do not permit the liquid to come in contact with bare skin for extended periods; use rubber gloves such as those worn by surgeons, or use appropriate tools. Don't let the liquid contact vulnerable plastics.

Step 3

Obtain a steel guitar, mandolin, or zither string slightly larger than the fiber diameter, and cut it with diagonal cutters, taking precautions not to permit the shorter bit to fly off and rebound. Bend the longer piece at right angles so that you can hold it easily, and place the cut end firmly in the hole left by the ball bearing. Maintain pressure, and roll the pen in your fingers while the music wire increases the size of the ink channel through the center of the tip. This will take only a few minutes. The wire may also be used to assist in cleanout of the tip and reservoir.

Step 4

When the tip is clean, insert the fiber into the reservoir, and extend it just beyond the tip, about the same distance at which the ball bearing previously extended. Mix a drop of each liquid of transparent epoxy five-minute cement. Place a tiny drop of the mix at the tip of the pen, while holding the fiber steady. Lay the assembly aside carefully for at least an hour so that the cement can catalyze thoroughly. Despite the advertised "five minute" catalysis time, tiny amounts take longer to set.

Step 5

Prepare a number of pieces of increasingly smaller spaghetti tubing to fit concentrically and progressively, each smaller piece extending about a centimeter longer than its larger counterpart. The final size should be about 2 cm longer at least, and if possible it would be best to extend it the full length of the fiber to form a protective sheath. Run the smaller spaghetti over the far end of the fiber, down into the pen, and inside the former ink reservoir if possible. Touch a drop of instant-drying cement to the entrance of the reservoir. (Obviously this is a very exacting and tedious operation; the hole in the pen top is always too small!) Give the cement 20 minutes to set, despite the "instant" advertising.

Step 6

Slip another, larger-diameter piece of spaghetti, cut shorter in length than the first piece, down into the pen, against the end of the ink reservoir.

Step 7

Slip the third, larger piece down over the reservoir as far as it will go. This will entail a cut-and-try procedure to get the piece the right length at the pen exit. This will give a substantial construction to your project, and provide excellent protection for the fiber at the exit, since the progressive cutting of the spaghetti lengths provides a larger bending radius at the fiber exit point.

NOTE: Plastic insulation from hookup wire will serve nicely for the first, long-length fiber-protective spaghetti. Removing the wire is not difficult. Place the bared end of the wire in a vise, and with a cloth or a glove to protect your hand, pull gently the full length of the insulated wire, letting the insulation slip through your glove or cloth. Reach quickly back to the beginning and do it again, over and over, the idea being to increase the heat level of the insulation at the same time that you urge it to slip down the wire in the direction of your stroke. After a few passes, the wire slips right out, to be replaced with the fiber.

Step 8

Prepare the coupling end according to the instructions in Project 3 (Chapter 17).

Using the Pen

A good first test for the pen is to run it across supermarket product-identification striations. These are generally black and white, and a fairly bright light shining on them will be reflected into the pen. You should hear a "z-z-z-ip" as you slip the pen across them.

This pen should be used in your finer searches for light activity, with the care and attention you would give a fine test instrument. It will give you some interesting effects and perhaps some ideas when held near the grooves of a phonograph record with a strong light in the vicinity. If the groove is illuminated with a laser diode, the results may surprise you.

Section 4
Appendixes

A. Caution Note
B. Light
C. Obtaining Experimental Fiber-Optics Supplies
D. Sources for Fiber-Optics Systems, Connectors, and Components
E. Building Circuits
F. Terminating and Coupling Optical Fibers
G. Examples of Manufactured Products
H. Miscellaneous Information
I. Advantages of Fiber Optics
J. Glossary
K. Bibliography

Appendix A: Caution Note

Take appropriate precautions when using hazardous materials. Warnings are given in the text as uses for such materials are suggested. The hazard may not be limited to yourself alone; a small child with access to your workbench may not be able to read warning labels and signs, or a pet could upset an open bottle of flammable liquid while you are temporarily out of the room. Always store hazardous materials out of the reach of children and pets, and of adults who are unfamiliar with the uses of your special materials.

Appendix B: Light

We are so accustomed to sight that we take light for granted, not generally realizing that, like radio waves, light is an electromagnetic phenomenon in the physical world. There is an infinite spectrum of such energy, extending from direct current (zero frequency) upward through "sound," radio, television, infrared, visible light, ultraviolet light, x-rays, gamma rays, and cosmic rays. We do not know of an upper limit of the spectrum, although for practical purposes, generation of upper-level frequencies in the laboratory falls short of cosmic rays.

"Light" is a very narrow band of frequencies along this spectrum, in the region of 10^{14} hertz (Fig. B-1). We most often refer to light by its wavelength rather than by its frequency, but either can be used. Wavelength may be derived at any point on our spectrum by dividing the speed of light, 300,000 km/s (186,000 mi/s) by the frequency at that point. Visible light extends from about 700 nanometers (nm) to about 400 nm.

The energy of light is an electromagnetic field that varies in strength in much the same way as the electricity in the power lines: sinusoidally. The rate at which it varies is far, far faster, with the power line at 60 hertz and the center of the visible light range in the billions of hertz. To refresh your understanding of the relationships inherent in a sine wave, we include Fig. B-2.

For a review of the prefixes used in fiber optics relative to numbers and their scientific notation, see Table B-1. There are two units of

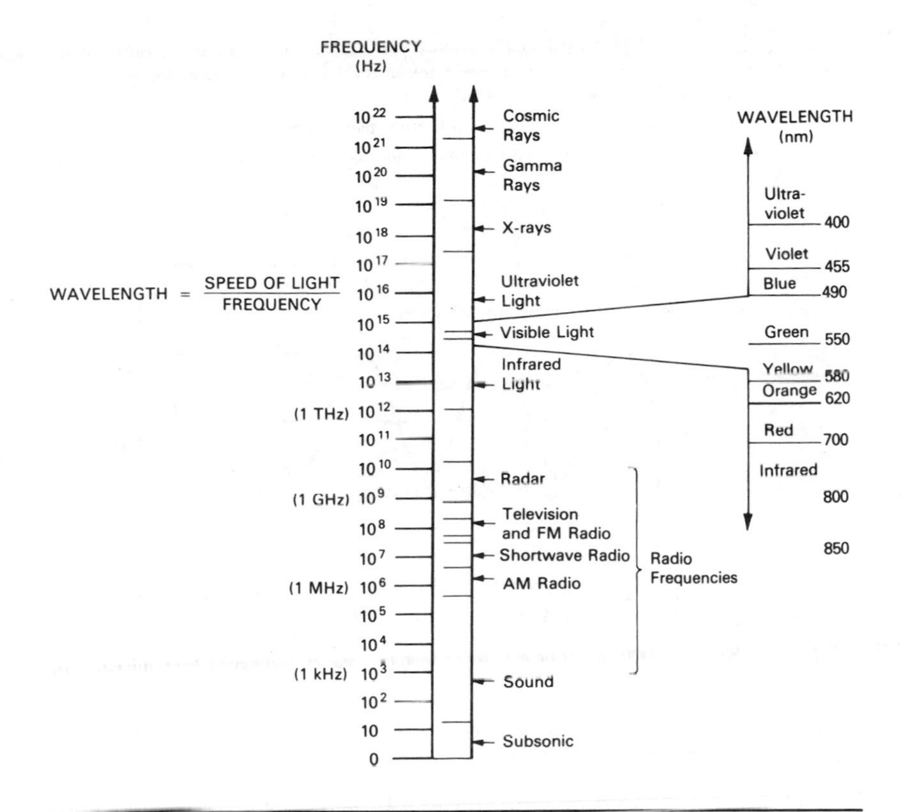

Fig. B-1. The electromagnetic spectrum. (Courtesy AMP Inc.)

measure that should be mentioned as well: 1 mil = 0.001 inch = 25 microns (approximately), and 1 micron = 10^{-6} meter = 1 micrometer (μm).

Infrared and ultraviolet light are both invisible to the human senses, although some animals and insects are known to have ranges of vision extending beyond the human range. Most fiber-optic links operate in the infrared range, because at this time components can be made to operate more efficiently at these wavelengths.

You may see various references in other books to microns (μ) and millimicrons (mμ) rather than to the international standard of scientific notation. In this book, we use the preferred "nanometers" (nm) to

A sine wave is one that rhythmically swings from a zero baseline up to a positive peak, down past the baseline to a negative peak, and then back to the baseline. Because the terms describing sine waves are essential, we offer a brief review.

Cycle is the complete swing through both the positive and negative alternations.

Frequency is the number of cycles completed in one second. The unit of frequency is the hertz (Hz). One Hz is equal to 1 cycle per second.

Wavelength (λ) is the distance between the same point on two consecutive waves; it is the distance a wave travels in a single cycle.

Amplitude is the height of an alternation; it is the distance from the baseline to either peak. The amplitude is a measure of the strength of the wave.

Period is the time it takes to complete one cycle.

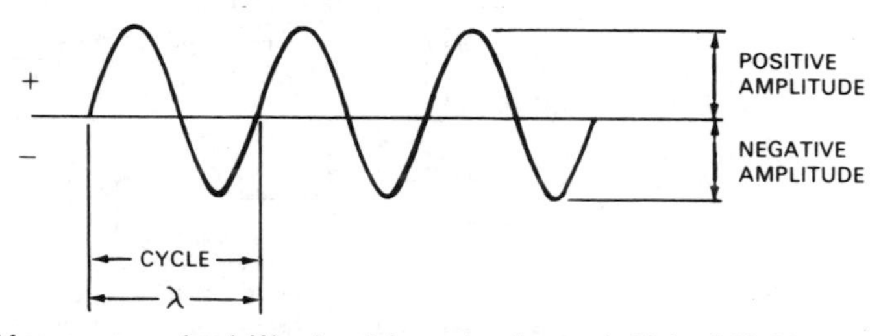

Wavelength and frequency are related. Wavelength is equal to the speed of light divided by the frequency. A 60-Hz wave has a wavelength of 3100 mi. A 55.25-MHz wave (the frequency that carries the picture for channel 2 on television) has a wavelength of 17.8 ft. By the time we reach the trillions of hertz of light, we are dealing with wavelengths of hundred-thousandths of an inch.

Fig. B-2. Sine-wave data. (Courtesy AMP Inc.)

express the wavelength of light, although each system is correct and each has its place. For example, 800 nm, 800 mμ, and 0.800 μ are equivalent.

It is relatively easy to express and understand the common numbers such as a few hundred, a few thousand, or even a million or two. But as we approach larger numbers into the billions and trillions, we tend to lose comprehension for their immensity. A similar vagueness is encountered as we extend into the world of the ultrasmall. It is not too great a stretch of our comprehension to encompass the idea of a unit being divided into one thousand parts. But when we think of a meter being divided into a million parts (a micrometer) and into a thousand million parts (nanometer), we begin to falter.

Working with very large and very small numbers is rather awkward until we express them in scientific notation. While noting them as

Table B-1. Scientific Notation

Prefix	Symbol	Multiplier	Power of 10	Denomination
Tera	*T*	*1,000,000,000,000*	10^{12}	*Trillion*
Giga	*G*	*1,000,000,000*	10^{9}	*Billion*
Mega	*M*	*1,000,000*	10^{6}	*Million*
Kilo	*k*	*1000*	10^{3}	*Thousand*
Milli	*m*	*0.001*	10^{-3}	*Thousandth*
Micro	*μ*	*0.000,001*	10^{-6}	*Millionth*
Nano	*n*	*0.000,000,001*	10^{-9}	*Billionth*
Pico	*p*	*0.000,000,000,001*	10^{-12}	*Trillionth*

powers of ten (scientific notation) makes manipulation easier, it does not immediately contribute to our comprehension of size, speed, or value. Yet as we begin to use scientific notation, we find our resistance to this form yielding to a better "feel" for the extremes that can be represented and easily manipulated.

Particularly in the field of light does this assist in our working with the almost unimaginably high frequencies and ultrashort wavelengths involved. Deep red light, for instance, has a wavelength of 700 nanometers, a convenient number. How long is 700 nanometers? If we divide a meter into a billion parts, deep red light spans only 700 of these miniscule units. A billionth of a meter is so tiny that the range of wavelengths of visible light—roughly 400 to 700 nanometers—makes little impression upon us.

On the other hand, wavelength has its equivalent in frequency, so we can say that deep red light has a frequency of 430,000,000,000,000 hertz. A number with that many zeros is beyond our ordinary comprehension. To say that the frequency of violet light is about 660,000,000,000,000 Hz is equally beyond ordinary comprehension. We tend to think of the difference between 430 and 660 and ignore the zeros, but this fails to convey the enormous spread of frequency between these two extremes.

If we give the numbers their international prefixes, as in Table B-1, and relate them to scientific notation, which involves the powers of ten, we find them easier to handle.

Appendix C: Obtaining Experimental Fiber-Optics Supplies

There are many "odd-lot" retail electronics dealers who advertise in electronics and amateur-radio magazines. These companies have thousands of parts bargains to offer, but while a certain purchase may be a "bonanza" on one particular month, it may be sold out when your order arrives.

You may buy a nice item on one order and find the product to be just what you wanted. But when you try to duplicate your buy a month later, it will be "temporarily out of stock" or "back-ordered." The "temporary" is most likely permanent, because at the next manufacturer's auction attended by that company's buyer, the factory will have something entirely different as an overrun from the assembly line. Therefore, it is impractical to specify, except in general terms, any parts or items from these sources to use in your experiments.

We mention parts obtainable from Radio Shack and G. C. Electronics, two of the major "walk-in" retailers, when these parts are listed in their catalogs. So far, however, such parts suppliers have offered only a token selection of fiber optics. Edmund Scientific lists a number of items for mail order. Technical Electronics Corp. has listed a 25-inch fiber-optics light pipe with ends prepared for mating with flat photodiodes and LEDs (Stock No. 1378).

Experimentation in fiber optics is active on many levels. From the home-workshop enthusiast to the most sophisticated government or industrial scientist, investigation into fiber-optics communications is from the simple to highly complex. It is too early to expect supply houses to have a wide range of inexpensive components with which to work, although heavy buyers and users obtain hundreds of thousands of units. Experimenters and students may have to make many of their own parts, but that is half the fun.

Ads for short bits of fiber of up to a meter or two in length can be found in electronic technical and hobby magazines. Longer runs of up to a thousand meters can be obtained by consulting the *Thomas Register* in your local library, although you may have to order through the manufacturer's agent (see also Appendix D). Occasionally, you may discover discarded display-type fiber optics in general-merchandise shops such as Goodwill Industries; sometimes these can be dismantled and the fiber strands salvaged for experimental use. These may or may not be coated fibers, and certainly you will not know whether they are step-index or graded-index fibers.

Light-emitting diodes are available in profusion from surplus and electronic-supply houses, mostly with but sometimes without spectrum data. Solid-state lasers will be more difficult to locate and much more expensive to purchase in single quantities.

A number of fiber-optics kits are offered by manufacturers for use by experimenters and engineers who want to evaluate fiber-optic performance. Two examples are shown in Figs. C-1 and C-2. The kit in Fig. C-1 contains sufficient components to house 25 emitters or detectors, terminate 20 fiber-optic cables, and make other types of splices. Its price at the time of writing is $165. Because of its diversity and relative expense, the experimenter's kit in Fig. C-2 is most likely to be purchased by companies that contemplate entering the fiber-optics field in some way. It contains 72 semiconductor devices, pc boards, connectors, and cables — enough to build six TTL or CMOS compatible fiber-optic data links. The price at the time of this writing is $350.

A brief source listing for various components and systems follows in Appendix D. This list should be used as a beginning, an indication of the kinds of places fiber-optics components are to be found.

By far the most satisfying and rewarding source of parts for the experimenter is his own ingenuity and inventiveness. Common items with

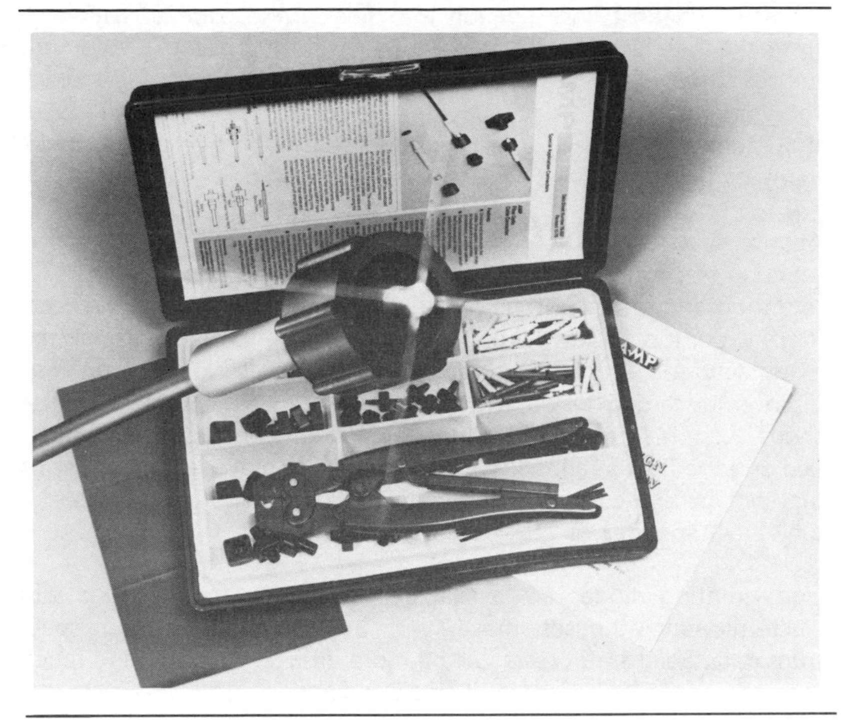

Fig. C-1. Kit of fiber-optic components. (Courtesy AMP Inc.)

intended purposes far removed from fiber optics can sometimes be pressed into service and will be found to work surprisingly well. The long-time experimenter will have accumulated a "junkbox" or series of shelves where hundreds of general items have been tucked away for some future use not foreseen at the time.

For instance, there are cadmium-sulfide elements in hand-held light meters that have gone out of style in the picture-taking field. Today, cameras are electronic and automatic, and outdated light meters are appearing in "as-is" store windows for a dollar or less. Old security systems sometimes contain some kind of photocell, from the vacuum-tube type to infrared sensors. The "rifle" used in some tv games contains a photodiode or phototransistor and lens system.

Serendipity (chance observation falling upon a receptive eye) can aid in locating parts and components for your fiber-optics experiments and projects. For instance, it may come as a surprise to users of LEDs

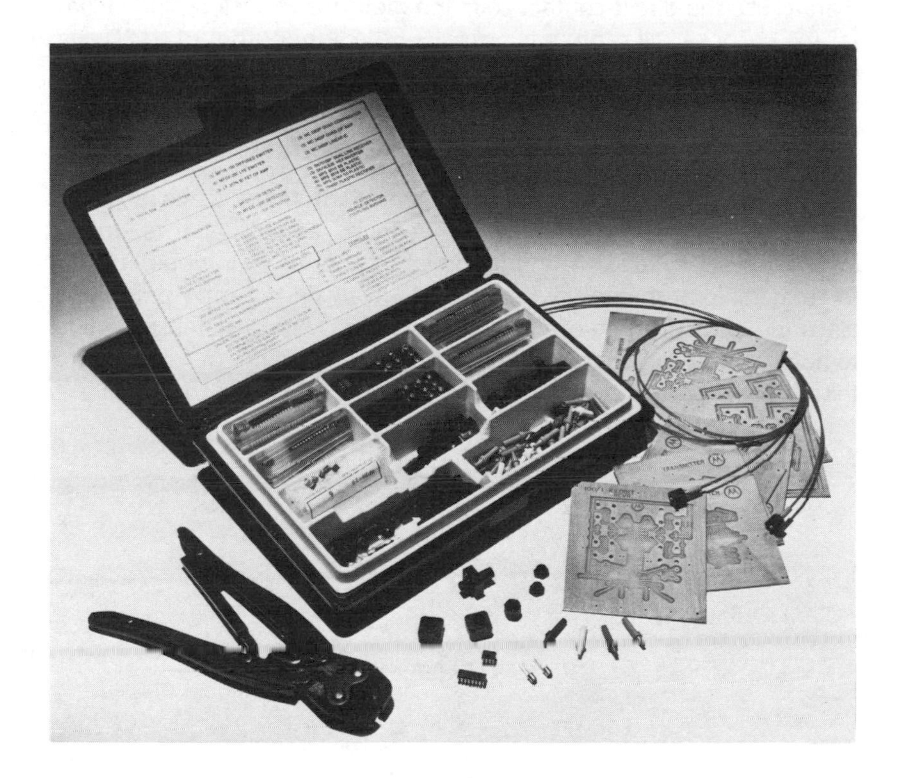

Fig. C-2. Experimenter's fiber-optic kit. (Courtesy AMP Inc.)

to learn that they may be used reversibly, that is, as photodiodes. A reverse bias (determined experimentally) is applied.

The same old 35-mm still camera or 8-mm movie camera that you cannibalize for the photocell will yield some interesting lenses, especially those used for viewfinding. These can be pressed into service for your fiber-optics experiments with only minimal changes in holders. The shutter mechanism in the movie camera will serve admirably as a motor-driven mechanical chopper. Cameras such as these, costing $250 or more a few years ago, now can be found for a dollar or two, and they are sometimes sold by the grab-bag full.

If selection or grinding of lenses is within your capabilities, you may want to use a lens to direct the light into the end of the fiber, rather than to use the direct-contact coupling methods. A well-designed lens for the purpose will capture almost all of the light output of a LED and place it into the cone of acceptance of the single-strand fiber. However, lens preparation is a most exacting task, and it is not recommended without considerable study.

Suppose you have located a small bundle of fibers, each about a meter long. Your experimental use requires at least a three-meter length of fiber to reach from the top left corner of your workbench to the underside at the right where it is relatively dark. Simply polish and place the ends of three of the fibers together, with a suitable binding to hold them. A purchased coupler to do this job can cost from two to thirty dollars. You can make one from wood or plastic for pennies.

Appendix D: Sources for Fiber-Optics Systems, Connectors, and Components

Sources

Several possible sources of fiber-optics hardware are listed in this section. For convenience, they have been grouped according to type of equipment or component. The addresses of these companies are given in the next section.

Connectors

AMP Inc.

Amphenol North America, Bunker Ramo Corporation

Augat, Inc.

Galileo Electro-Optics Corp.

ITT Electro-Optical Div.

Math Associates, Inc.

Cables and Systems

Math Associates, Inc.

Valtec

Handheld Scanners

Welch Allyn

Miniature Vacuum Lamps

Welch Allyn

Light Pipes

Welch Allyn

Fiber-Optic Bundles

Welch Allyn

Plastic-Fiber Cables and Bundles

Poly-Optical Products Inc.

Plastic Assemblies

Texas Instruments Inc.

Systems, Components, and Accessories

Augat Inc.

Burr-Brown

Galileo Electro-Optics Corp.

Fiber, Lenses, and Experimenter's Kits

Edmund Scientific

General Electric Co.

Heath Company

Math Associates, Inc.

LEDs, Phototransistors, Miscellaneous Parts

Technical Electronics Corp.

Company Names and Addresses

AMP Inc.
449 Eisenhower Blvd.
Harrisburg, PA 17105

Amphenol North America Div., Bunker Ramo Corp.
2122 York Rd.
Oak Brook, IL 60521

Augat, Inc.
33 Perry Ave.
P.O. Box 779
Attleboro, MA 02703

Burr-Brown
International Airport Industrial Park
P.O. Box 11400
Tucson, AZ 85734

Edmund Scientific
101 E. Gloucester Pike
Barrington, NJ 08007

Galileo Electro-Optics Corp.
Galileo Park
Sturbridge, MA 01518

General Electric Co.
101 Merritt 7
P.O. Box 5900
Norwalk, CT 06856

Heath Co.
Benton Harbor, MI 49022

ITT Electro-Optical Div.
Plantation Rd.
Roanoke, VA 24019

Math Associates, Inc.
376 Great Neck Rd.
Great Neck, NY 11021

Poly-Optical Products Inc.
1815 E. Carnegie Ave.
Santa Ana, CA 92705

Technical Electronics Corp.
27 Gill St.
Woburn, MA 01801

Texas Instruments Inc.
P.O. Box 225012 M/S 308
Dallas, TX 75265

Valtec
West Boylston, MA 01583

Welch Allyn
Jordan Rd.
Skaneateles Falls, NY 13153

Walk-In Retail Stores

G.C. Electronics
Independent Retail Stores
Radio Shack

Mail and Telephone Order via UPS and USPS

Ancrona
P.O. Box 2208
Culver City, CA 90230
Tel 213-641-4064

Digi-Key Corp.
Tel 800-346-5144

Fuji-Svea
P.O. Box 40325
Cincinnati, OH 45240
Tel 800-421-2841

Jameco Electronics
1355 Shoreway Rd.
Belmont, CA 94002
Tel 415-592-8097
($10.00 minimum order)

Meshna Inc.
P.O. Box 62
E. Lynn, MA 01904

Solid State Sales
P.O. Box 74
Somerville, MA 02143
Tel 617-547-7053

Published Lists of Companies

For an up-to-date list of all companies that offer fiber-optics products, check any large technical library for:

Microwaves and Laser Technology
September 1981 Edition
Hayden Publishing Co.
50 Essex St.
Rochelle Park, NJ 07662

The Optical Industry and Systems Purchasing Directory
Optical Publishing Co., Inc.
Box 1146 Berkshire Common
Pittsfield, MA 01201

Sample Data

Table D-1 lists some examples of fiber-optics prices, and Table D-2 lists some of the technical parameters for selected fiber cables. These tables have been included only to provide a general idea of the prices and specifications that could be expected at the time this book was being prepared. They do not necessarily imply an endorsement of any of the products, and, of course, there may have been subsequent changes in some of the information.

Table D-1. Price Examples, Fiber-Optics Manufactured Products*

Item	Description	Price
XA-1000 Analog Transmitter	Dc – 3.5 MHz	$195.00
RA-1000 Analog Receiver	Dc – 3.5 MHz	$235.00
XA-1800 Audio Transmitter	10 Hz – 40 kHz	$130.00
RA-1800 Audio Receiver	10 Hz – 40 kHz	$130.00
XD-1100 Digital Transmitter	0 – 100,000 pps	$125.00
RD-1100 Digital Receiver	0 – 100,000 pps	$125.00
AOF 1040 Optical Fiber	0.3-mm Quartz	$1.70/meter
AOF 1300 Optical Fiber	1.0-mm Plastic	$1.50/meter

*1980, Math Associates, Inc.

Table D-2. Fiber-Optics Cable Parameters for Selected Products

Mfr.	Type	Attenuation (dB/km)	Wavelength (nm)	Bandwidth (MHz/km)
Belden	225000	10	820	200
	220000	10	850	25
	221000	10	850	20
Corning	5040/41	5	820/900	400
	10020/21	10	820/900	200
DuPont	PFX-S120	40	775	
	PFX-P140	385	650	
	PFX-P1R	320	690	
Galileo	3000D-LC	50	850	
	3000D-7	100	850	
	3000D-19	100	850	
ITT	T-200	5	850	3.5 ns*
	T-320	10	800	30 ns*
	T-300	10	800	30 ns*
	T-100	3	1060	15 ns*
Valtec	XDMG05-06	5	820	400
	HDPC10-02	15	820	10
	LDSG04-01	10	820	1 ns*

*ns = nanoseconds square wave or pulse

Appendix E: Building Circuits

The solderless breadboard is particularly suited to building experimental circuits. Changes in parts placement may be made rapidly and without endangering heat-sensitive items. Troubleshooting is made easier because a misconnection can be corrected by unplugging a lead or jumper and inserting it into the correct pin jack.

The plastic modular boards are excellent. Examples are the Radio Shack 276-174 and the Vector 51X Klip-Blok/Strip breadboarding system. These provide rows of preconnected pin jacks to take discrete transistors or DIP chips. Snap rails are provided for attaching two or more boards together and for attaching a small panel punched to take a potentiometer, switch, binding posts, and other hardware.

When all the bugs have been corrected in a given experimental setup on a board, you may wish to transfer it to more permanent form. For this purpose, there are pre-etched solderable pc boards with the identical patterns found on the solderless boards. Thus, you can maintain the same parts relationship on the permanent project without having to etch your own special board.

Soldering is not the only way to secure connections on printed circuit boards or on prepunched boards without circuit prints. Wire wrapping around pins or leads is quite acceptable when performed with a regular tool. One such tool is Radio Shack's 276-1570.

A combination of soldering and wire wrapping can be useful, especially for mounting the longer-terminal DIP sockets. Solder the terminals to the pc board, and leave the long leads for use with the wire-wrap tool. Prepunched, nonprinted boards take to wire-wrap methods quite well; resistor, capacitor, and other parts leads are left long enough beneath the board to accommodate the hand-held tool and wire.

The longer leads of parts can be inserted into the prepunched board, bent in the direction of connection to the next component in the circuit, and soldered directly to that point. The leads can be fashioned to form the same pattern that would exist had the circuit been printed and etched. Jumpers may be inserted on top of the board to complete circuits that otherwise would have to cross over a previous circuit connection. Crossovers should be avoided, but when one is absolutely necessary, adequate insulation should be provided in the form of spaghetti or a wafer of thin insulating material.

Appendix F: Terminating and Coupling Optical Fibers

However long a fiber may be, eventually it must be terminated in a manner that permits it to function as a light carrier from both ends. Ideally, no rays of light approaching its acceptance angle would be lost by reflection or refraction, and no rays of light having traveled its length would be reflected or refracted back into it and toward the origin of the light. Further, no rays would be refracted out its far end at an angle greater than the calculated divergence angle.

In laboratory testing of fiber ends, a helium-neon laser is directed into one end of a prepared fiber, and the other end is placed perpendicular to and within 3 to 5 cm (1 to 2 inches) of a white screen in a darkened room. A well designed fiber will demonstrate a clean, circular output pattern of light with a cone of divergence that very closely mirrors the cone of acceptance of that particular fiber.

Cleaving the Fiber

Such testing has demonstrated cleavage techniques that produce virtually perfect cones of acceptance and divergence without polishing. We do not mean to imply that cleaving such small-diameter objects is easy and simple. Care and attention to detail of a nature that would do credit to a watchmaker are necessary. With such care and attention, a fiber may be cleaved with a tiny triangular needle file that is, incidentally, one of the common hand tools of the watchmaker's trade (Fig. F-1A). Holding the fiber between the thumb and forefinger of each hand, lower it gently over the file, rotate the fiber 360°, and lift it off the file. Do not try to rotate and scratch in the reverse direction, because a double scratch will result. Some technicians use a single scratch only, without rotating the fiber.

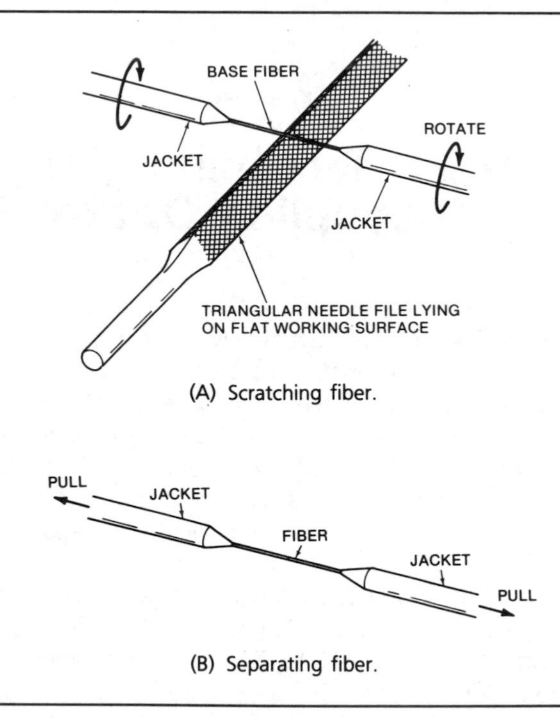

Fig. F-1. Method of cleaving optical fiber.

A constant, even tension applied from both directions away from the microscopic scratch will cause the fiber to break cleanly around the scribe line, in much the same manner that a glazier scribes and breaks a pane of window glass (Fig. F-1B). If this is carefully done, no further polishing will be necessary. Fiber diameters as small as 50 μm have been cleaved in this way with good results.

Such cleaving is done routinely in industry, with a sapphire mounted in a specially designed cleaving tool used instead of a needle file. Commercially available sapphire-mounted tools are very expensive and can be justified only if used daily over long periods by competent employees. However, a determined, innovative experimenter might find the mounted diamond in an engagement or lodge ring to be exactly the right tool for the job. Innovation will remind the experimenter that sapphire and diamond scribers are within easy reach of the high-fidelity music buff; discarded phonograph needles, sharpened from their original oval shape by playing too many records, can make excellent scribers. The search for a suitable tool can extend into the local machine shop, where a chipped tungsten-carbide lathe tool might be found. Carbide is next to diamond in hardness, and being brittle, it chips quite easily; a carbide chip is very hard and very sharp, admirable qualities for a fiber-scribing tool. A good quality razor blade has been used with modest success, although the blade would require honing to be used thereafter for its original purpose.

The most advisable route to fiber-end preparation is to buy a commercial finishing kit that contains the necessary abrasives, holders, and flat working surfaces. For preparation of plastic fibers, fine emery paper may be used, followed by commercial silver-polishing compound and gentle rubbing on fine felt. Plastic surfaces are soft and easily scratched, so the materials in use must be clean and pure to avoid destroying considerable labor in a single stroke. A smooth-surfaced leather belt may prove to be useful in a polishing process.

There is no doubt that a microscope would be very convenient for observing the progress in polishing a fiber end, but few experimenters can afford to own one. Perhaps your local high-school or college lab will make one available for intermittent use. Another tool that is almost as useful as a microscope is a jeweler's loupe; these can be obtained quite inexpensively from Edmund Scientific. Don't overlook the lens from an 8-mm movie projector. Used with the light traveling the same direction as it does in projecting film, this lens held to the eye makes an excellent magnifier (but a poor loupe because of its weight). It is all too easy to remove from the projector, as many owners of these rapidly disappearing machines will complain. Old projectors can be obtained from Goodwill Industries stores and other outlets for no more than the cost of a good loupe, and they contain a number of other useful parts for the "junk box," including the condenser lenses, concave mirror, and high-intensity light source. The viewfinder lens of a movie camera makes a potentially useful lens system for fiber-optics and laser experimentation. Movie cameras are likewise being discarded since the advent of video recording.

The procedure for cleaving a fiber begins with cleaning.

A. Remove the jacketing materials, grease, and dirt from about an inch on either side of the intended cleavage point.
B. With the fiber firmly held between the thumb and forefinger of each hand, as close to the cleavage point as possible, apply a firm (but not excessive) tension.
C. Gently lower the fiber against the cleaving tool, and cause it to make a complete 360° turn (Fig. F-1A). Only the slightest scratch is needed; remember the size of the material you are working with. A simple scratch or nick on one side of the fiber may be all that is necessary, rather than trying to rotate the strand.
D. Increase tension on the fiber until it parts (Fig. F-1B). It will part at the scratch, and if the effort has been successful, the ends will be virtually as "polished" as though you had used the finest polishing procedures.

Plastic fibers may be cut in the same manner, but you might try a slightly different approach. Lay the cleaned portion of the fiber on a hard, flat surface, and grasp it between your thumb and index finger while you apply gentle but firm pressure at the point of cleavage with a razor blade or hobby knife. Roll the fiber 360° between your finger and thumb while applying this pressure, and the fiber will "fall apart" at the working point.

Do not use diagonal cutters, since these mangle and split the fiber longitudinally. Seen through a microscope, the cutting edges of "dikes" are about as sharp as a round file.

Rough cutting of fiber bundles is not quite the relatively simple matter described above. Unless the fibers are firmly bound with a cement such as epoxy, trying to cleave the bundle will leave each individual fiber a different length. Further, roughing and polishing of the fiber ends will be a longer, more arduous process unless the bundle is made rigid.

Instant cement will work well in making the bundle rigid. A tiny drop will enter the bundle by capillary attraction, and within a few minutes the end will be ready for polishing (catalyzing time is not quite the advertised "instant"). Epoxy cement will work well also, but it will take longer to catalyze (harden).

When the bundle is rigid, a diamond-dust–impregnated jeweler's saw would be excellent for scribing around the cable for cleavage, but since these are available only at a very large cost, the experimenter will do well to innovate. A triangular needle file will do a passable job, as will a good-quality razor blade if used on plastic fibers. Again, a straight, firm pull at the point of marking is recommended, although you will find that the bundle is somewhat stronger than you might at first anticipate. The cable will require grinding and polishing, because scribing individual fibers of a bundle is out of the question.

Grinding and Polishing

A tool such as a machinist's pin vise would be very handy for holding the fiber or bundle of fibers for grinding and polishing (Fig. F-2). Such a tool may be fashioned from a multimeter test prod designed to hold the old-type phonograph needle. Some kind of fixture must be used, because trying to make a perpendicular, flat grind on the end of a semiflexible cable by guess and feel is to invite frustration.

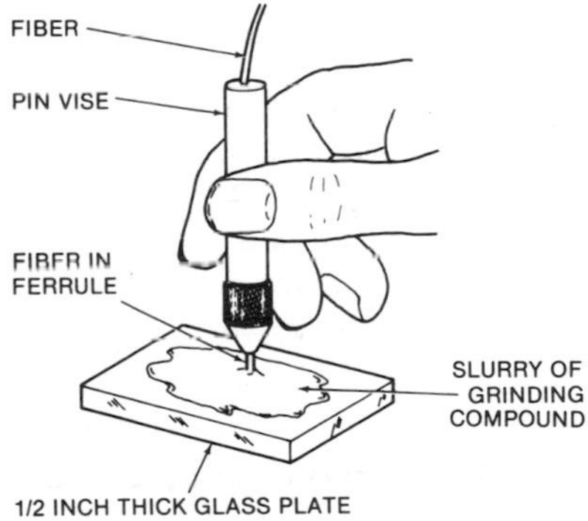

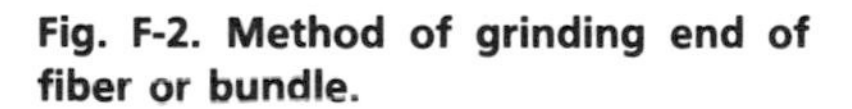
Fig. F-2. Method of grinding end of fiber or bundle.

There is a certain kind of pencil to be found in stationery stores that takes a very thin lead and feeds the lead by pressing a button at the pen top. This pressure opens a collet and permits the lead to slip downward a slight amount. One brand name for this type of pen is "Pentel," although others are on the market. Such a pen should make an excellent polishing-tool holder for small fibers.

An old ball-point pen or felt-tip pen of the fine-point type may be fashioned into a holder.

Once the holder problem is solved, rough grinding can begin:

A. Hold the fiber vertically while moving it in figure-eight and circular patterns through an alcohol/water slurry of No. 400 aluminum oxide, using a thick piece of glass for the anvil (Fig. F-2). Do not apply pressure; the weight of the holder is sufficient. Another satisfactory abrasive is 400-grit emery paper, but few hardware or supply stores carry such paper. Various grades of aluminum oxide can be obtained more easily. Edmund Scientific (Appendix D) is one possible source of supply.
B. Rotate the holder every five strokes, but keep in mind that there is a limit to rotating in the same direction without snapping the fiber. Rotate about 90°, start the other way, go for about 180°, reverse direction, then go 270°, and reverse again. This will avoid excessive twisting of the fiber.
C. When the end looks fairly smooth, wash the grit from the fiber, the holder, and the anvil. This must be done thoroughly and carefully, because any coarse grit left when you change to a finer polish will create havoc at the surface.
D. Repeat the above processes using No. 600 aluminum oxide. About 25 strokes will be required for glass; fewer should suffice for plastic.
E. Thoroughly wash the grit from the tool, the fiber, and the anvil. Attach a piece of surgical adhesive tape to the anvil, and place a drop of cerium-oxide – alcohol mixture on the tape at one end. Gently move the tool in this mixture with the same stroking as called for above, for about 30 to 40 strokes, gradually moving into the dry area of the tape. Remember: apply no downward pressure on the holder; let the weight alone do the work.
F. Wash the fiber with alcohol, and inspect the polished end. It should be bright and shiny to the eye, and under a microscope, if available, it should present a quite smooth surface. If any scratches are evident on the surface, repeat the steps from the beginning.

Plastic fibers may be polished as above, but because of their softer nature, each step should be adjusted downward in the total number of strokes.

These procedures are suggested only as a starting point. Creative experimenters will develop methods of their own, perhaps superior to these. These steps, when followed, will produce workable results.

Manufactured Couplings

Among the most comprehensive and helpful manuals available on a limited request basis is AMP Incorporated Handbook HB 5444. This handbook describes in considerable detail, among other things about fiber optics, the manner in which couplings are made and the requirements which dictate the manufacturing practices. A number of their illustrations have, with their permission, been reproduced in this book. Figs. F-3 through F-15 are representative of the illustrations in the handbook that deal with couplings.

Fig. F-3 illustrates one method of connector design that consists of a pair of concentric tubes. These tubes must have very close-tolerance manufacturing to provide control of fiber separation.

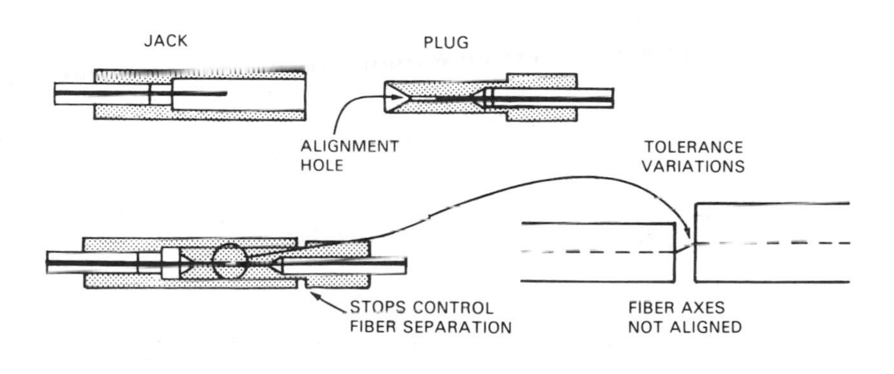

Fig. F-3. The tube-alignment connector method. (Courtesy AMP Inc.)

The straight-sleeve connector in Fig. F-4 uses a straight precision sleeve to mate two plugs; a nut controls fiber end separation. The double-eccentric connector in Fig. F-5 provides a method of controlling fiber end alignment. It consists of two separate eccentric sleeves that can be rotated individually to achieve a match of the fiber ends. The tapered-sleeve method of matching is illustrated in Fig. F-6. It uses a tapered plug and sleeve with compliant plastic buttons that squeeze together to bring the fibers into a good match.

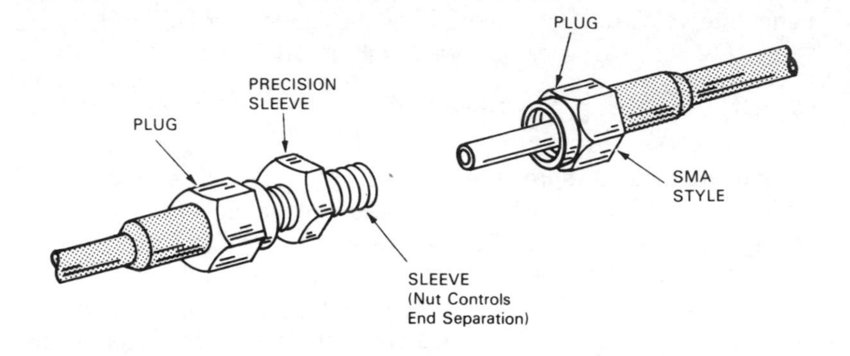

Fig. F-4. A straight-sleeve connector. (Courtesy AMP Inc.)

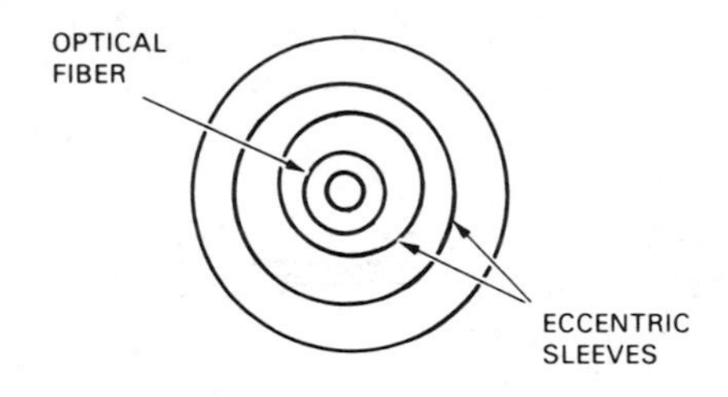

Fig. F-5. A double-eccentric connector. (Courtesy AMP Inc.)

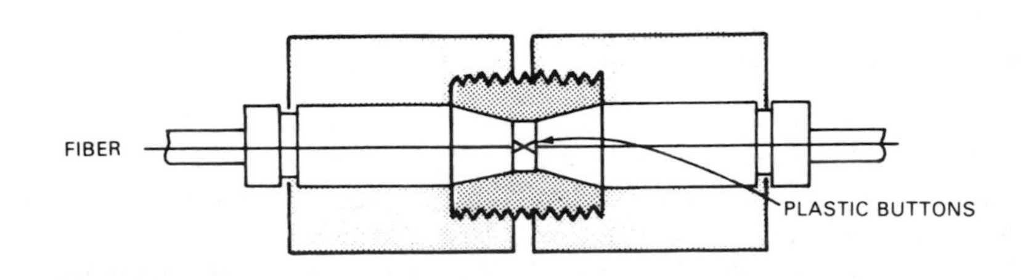

Fig. F-6. The tapered-sleeve coupling method. (Courtesy AMP Inc.)

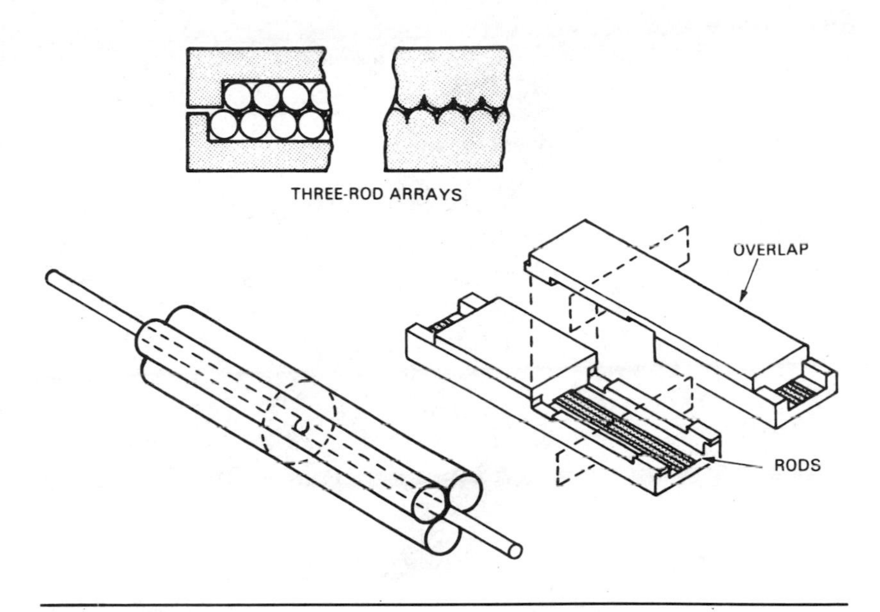

Fig. F-7. Three-rod connector. (Courtesy AMP Inc.)

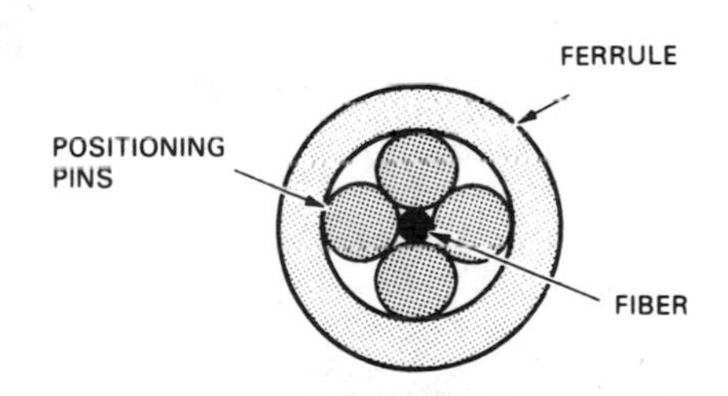

Fig. F-8. Four-pin coupling method. (Courtesy AMP Inc.)

The three-rod method of connector design is well noted for its success in matching fiber ends very closely. By using rods in rows, this method can be expanded to hold multiple fibers for coupling, as illustrated in Fig. F-7. This method may be extended into the four-pin design (Fig. F-8), generally held within a ferrule.

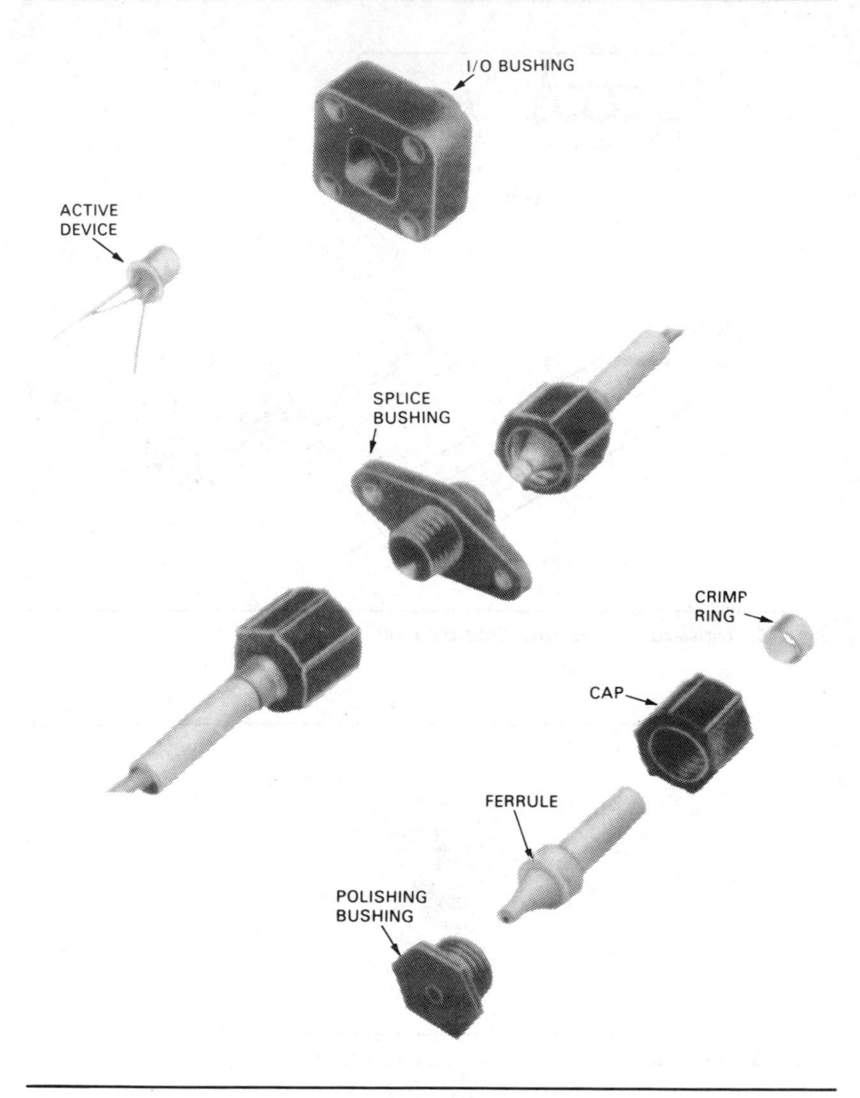

Fig. F-9. Single-Position connector. (Courtesy AMP Inc.)

The design of the AMP Single-Position connector is illustrated in Fig. F-9. This connector comes with a polishing bushing that is discarded after the polishing operation. This is a handy device, in that no more

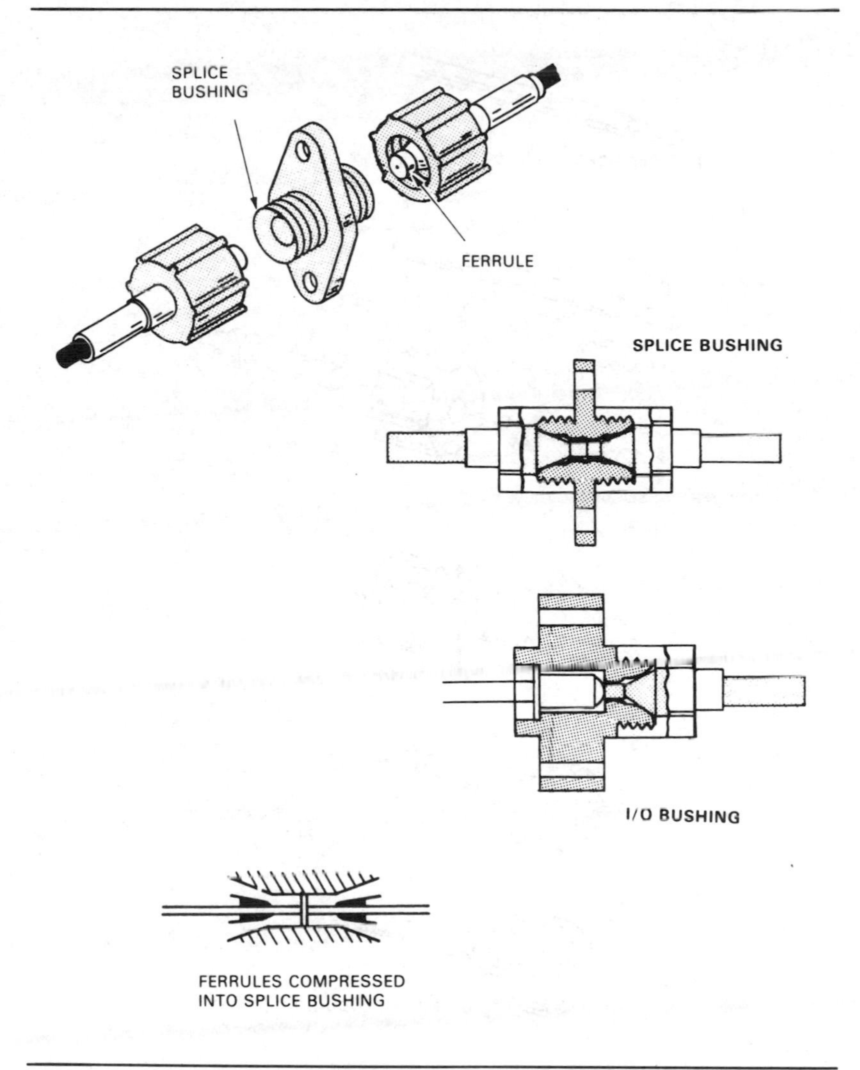

Fig. F-10. The AMP resilient-ferrule alignment mechanism. (Courtesy AMP Inc.)

fiber needs to be bared than exactly what is needed to fit into the bushing. The ferrule is sealed by the compression of the cap when screwed on, the ferrule itself being the resilient alignment mechanism, as shown in Fig. F-10.

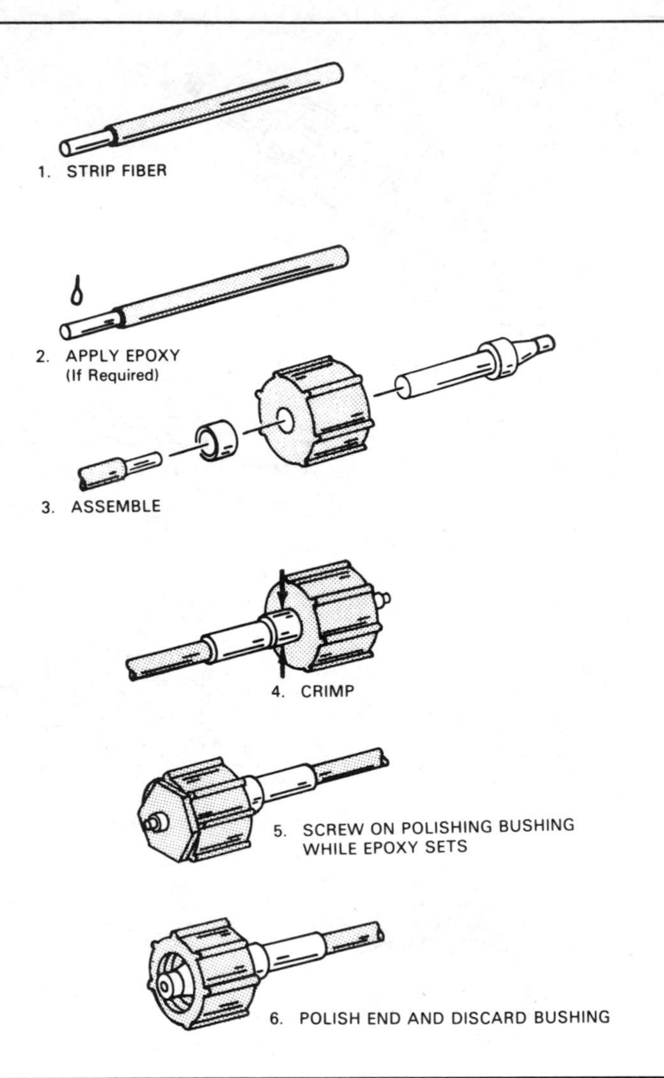

Fig. F-11. The AMP recommended termination sequence. (Courtesy AMP Inc.)

The fiber termination sequence recommended by AMP (Fig. F-11) begins with stripping the fiber and includes crimping with the tool supplied with their kit. Epoxy is applied, if required for the fiber used, and the polishing bushing is used as the guide for hardening.

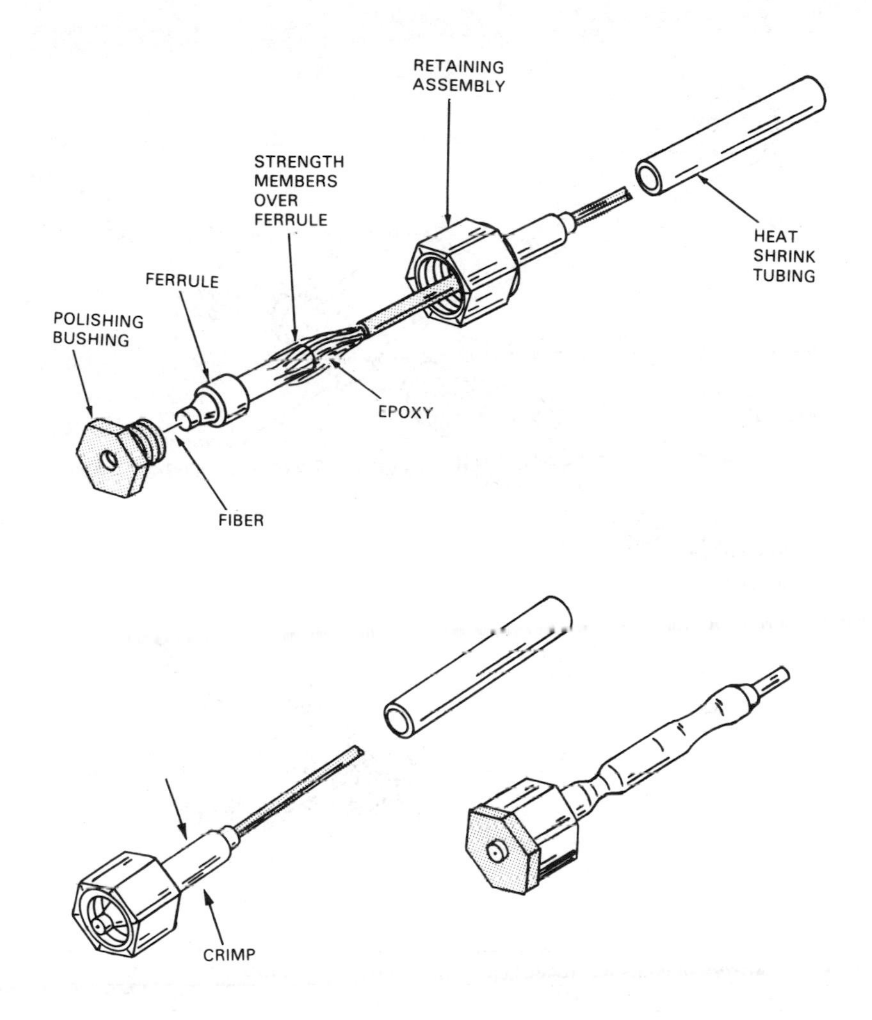

Fig. F-12. Typical termination of Small-Fiber connector. (Courtesy AMP Inc.)

The Small-Fiber connector is illustrated in Fig. F-12. The ferrule and metal retaining sleeve of this unit must be purchased to fit the particular cable to be used.

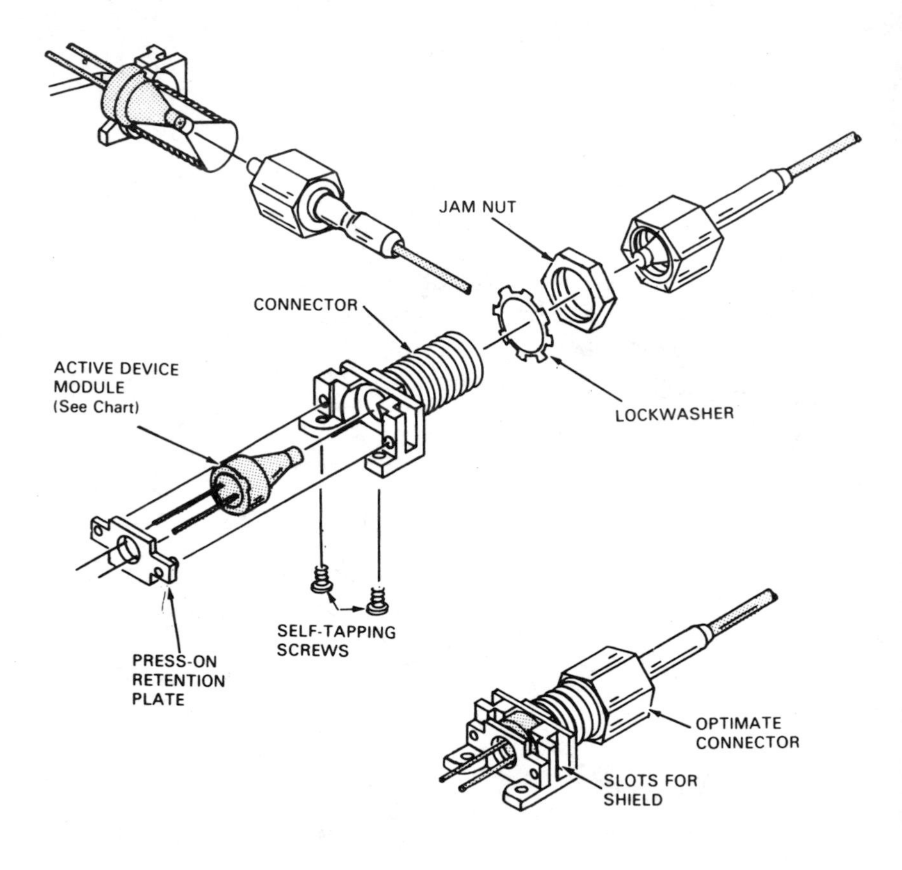

ACTIVE DEVICES AVAILABLE FROM MOTOROLA SEMICONDUCTOR PRODUCTS				
DESIGNATION	DESCRIPTION	RESPONSIVITY	RISE TIME	POWER OUTPUT
MFOE 102F	LED	—	25 ns	70 μW @ 50 mA
MFOD 102F	PIN Photodiode	.2 μA/μW	1 ns	—
MFOD 202F	Phototransistor	100 μA/μW	2.5 μs	—
MFOD 302F	Photodarlington	2.8 mA/μW	40 μs	—
MFOD 402F	Integrated Detector-Pre-amplifier	1 mv/μW	20 ns	—

Fig. F-13. AMP Active Device connector. (Courtesy AMP Inc.)

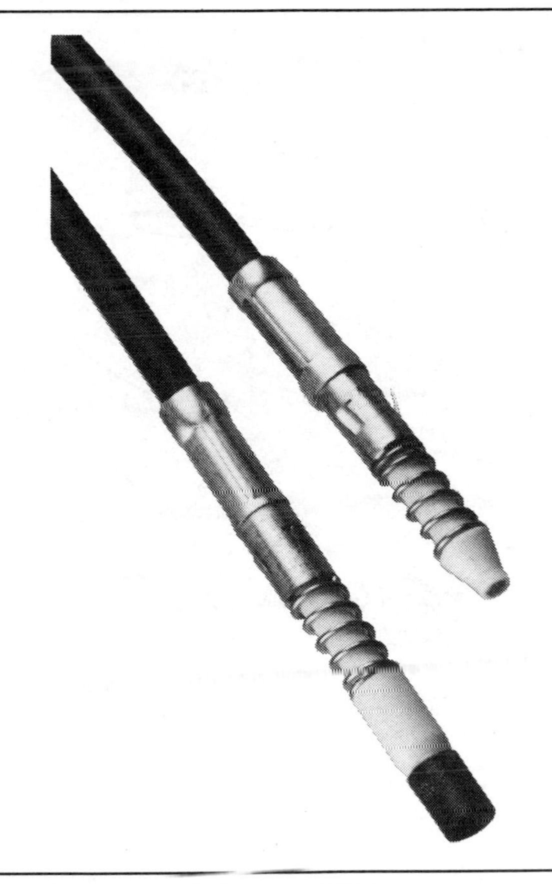

Fig. F-14. Mating ferrule pair. (Courtesy AMP Inc.)

The Active Device connector (Fig. F-13) permits the connection of an active device such as a phototransistor or LED to the Single-Position connector or the Small-Fiber connector previously mentioned. The active devices recommended for use are listed in Fig. F-13.

Fig. F-14 shows couplings that fit together with compression springs to maintain a tight coupling between fiber ends. This mating ferrule pair snaps into over 100 different cable connectors. The pair is self-aligning and is sized to mix well with electrical power, signal, and coaxial conductors in the same connector.

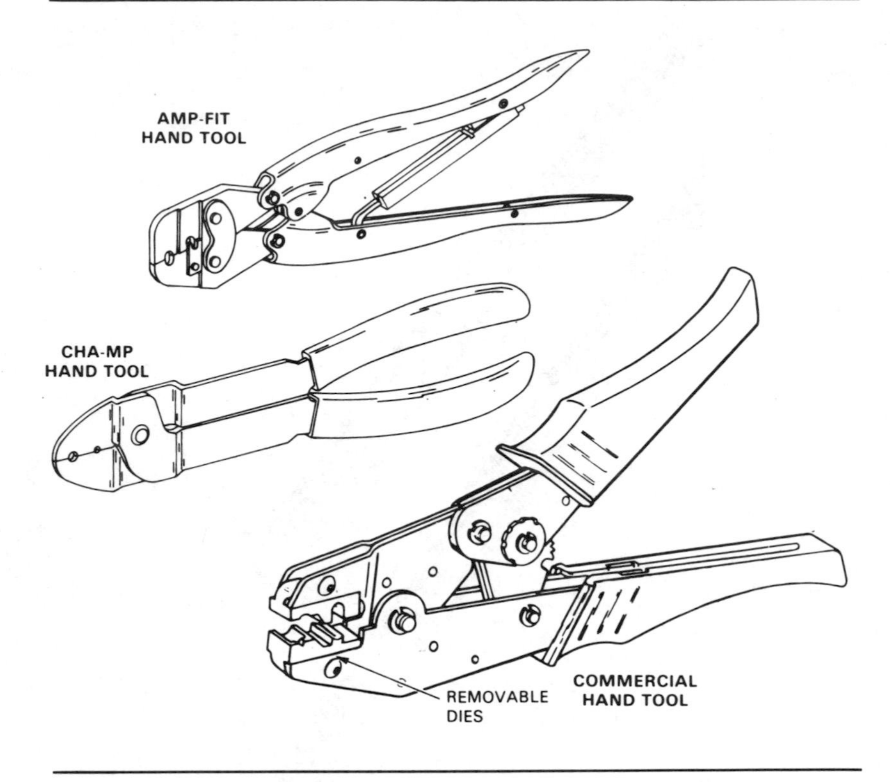

Fig. F-15. Crimping tools. (Courtesy AMP Inc.)

Crimping tools such as AMP's three models (Fig. F-15) are a must for use with their couplings, but such tools will find considerable general use around the experimenter's workbench in addition. They can be used to crimp television concentric-cable connectors, which, as you discovered in Projects 1 and 2, make excellent fiber-optics couplings.

Appendix G: Examples of Manufactured Products

Opto Technology, Inc., of Wheeling, Illinois, manufactures a fiber-link transmitter-receiver pair typical of the standardization entering the fiber-optics market. Their trademarked OP-TEC LINK OTD-500 fiber-optic data link consists of a transmitter unit and a receiver unit, each only a half inch in height, constructed in a 24-pin DIP package with pins ready for soldering into a circuit board. A male socket fitting for a fiber-cable connector is mounted at one end. The small size of the units is evident from the outline drawing in Fig. G-1A.

The transmitter takes current from a 5-volt dc regulated supply at pin 24; a TTL-compatible pulse signal is applied at pin 5. A ground, or common, return connection at pin 12 completes the circuitry. The device uses a GaAs (infrared) LED emitter to initiate the light beam.

The receiver also requires a regulated 5-volt supply. A potentiometer for adjustment of sensitivity is connected between pins 3 and 7; its resistance range depends on the length of the fiber link. A 5- to 10-foot link takes a 1-megohm potentiometer, whereas a 50-foot link takes a 1-megohm resistor in series with the 1-megohm potentiometer.

The system electrical characteristics include a receiver output signal rise and fall time of 1.0 microsecond, measured to ground, at a bit rate up to 50 kilobaud per second. By connecting the output to a 74LS14 Schmitt trigger chip, the rise and fall times can be reduced to about 20 nanoseconds.

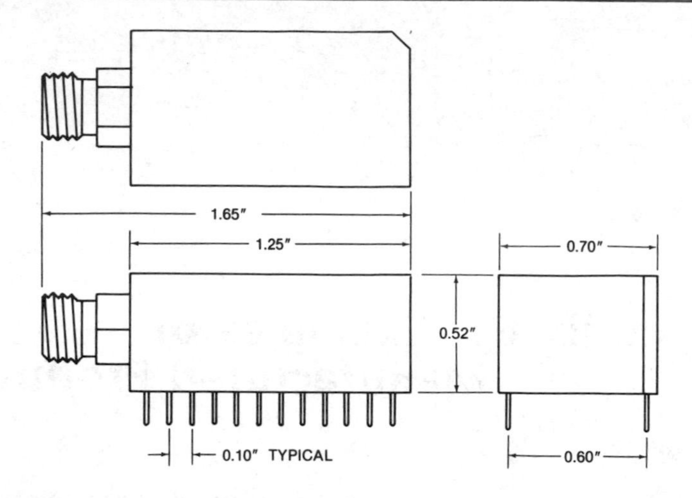

(A) Package outline.

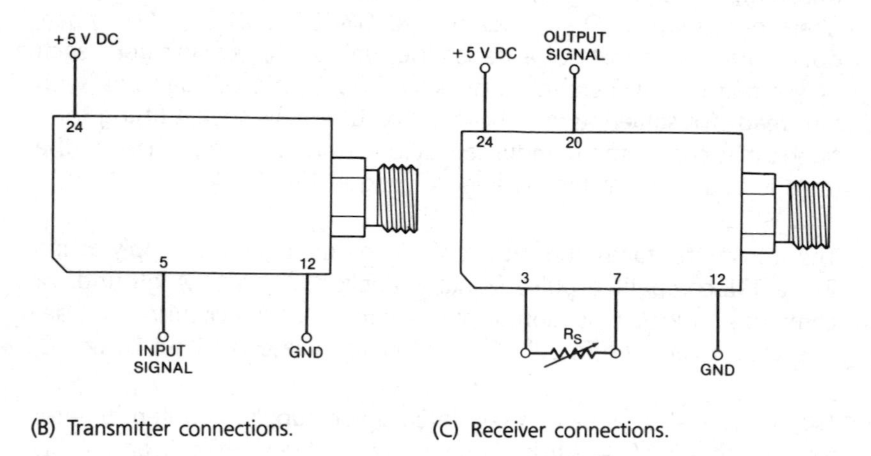

(B) Transmitter connections.

(C) Receiver connections.

Fig. G-1. OP-TEC transmitter and receiver.

This system can be used for audio, video, and computer communications. The receiver is featured as a "low-cost" fiber-optic data unit of hybrid construction. By "hybrid" is meant that a "chip" amplifier is incorporated in the package.

Opto Technology also makes packaged interrupters, optical switches that permit noncontact switching up to 10,000 operations per second. These are used in manufacturing operations for counting items leaving a production line, for instance. They would make excellent security devices as well, as burglar-alarm sensors.

An interesting and potentially useful Opto Technology product is the reflective transducer. This is a tiny unit that contains both a transmitter and a receiver/amplifier and that senses a reflective item at its output/input end. Suggested applications include sensing limits such as the end or beginning of a tape in a tape recorder, and reading printed code from a strip of paper tape. When the device is used with an infrared filter, ambient illumination effects are reduced virtually to nil, so it can be used in lighted areas without giving spurious results.

Edge-punched cards and punched paper tapes may be read by Opto Technology's 9-channel optical readers. Each channel contains a LED light source, a photosensor, a level detector, and a Schmitt trigger.

Appendix H: Miscellaneous Information

Pulses

This discussion of pulses is presented courtesy of AMP Inc. Fiber optics is well suited for digital applications, which means pulses. A pulse is a sudden burst of energy. In fiber optics, a pulse is formed by turning the source on for a brief instant. The burst of light is the pulse. Digital implies two states: on/off, 1/0, high/low. These two states represent bits, or binary digits. Not only does the burst of light have significance (1), its absence has significance (0); see Fig. H-1. A train of pulses or bits is given meaning through any of a variety of coding schemes.

The following are some of the important terms associated with pulses:

Amplitude is the height of the pulse; it is a measure of the strength of the pulse.
Width is the time the pulse remains at its full amplitude.
Rise time is the time to go from 10% to 90% of the amplitude. In fiber optics, it is related to how long it takes the source to turn fully on.
Fall time is the time it takes to go from 90% to 10% of the amplitude. It is related to how long it takes the source to turn fully off.

Rise time is perhaps the single most important characteristic in high-speed digital applications, for it determines how many pulses per second are possible. In the two pulse trains in Fig. H-2, the time slots and the pulse widths are identical. The only thing that allows more pulses in one train is shorter rise and fall times.

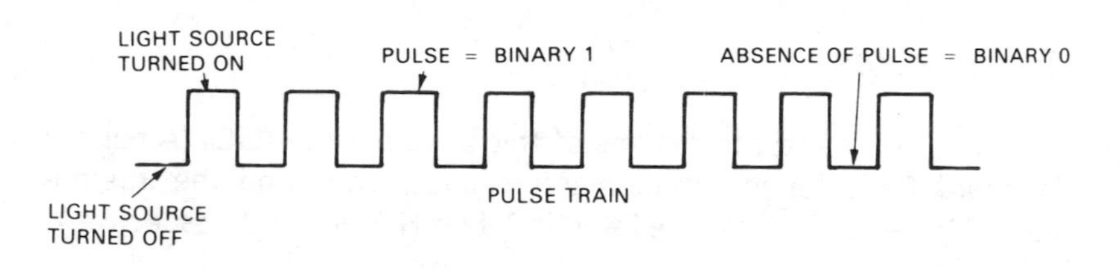

Fig. H-1. Two states of a pulse train. (Courtesy AMP Inc.)

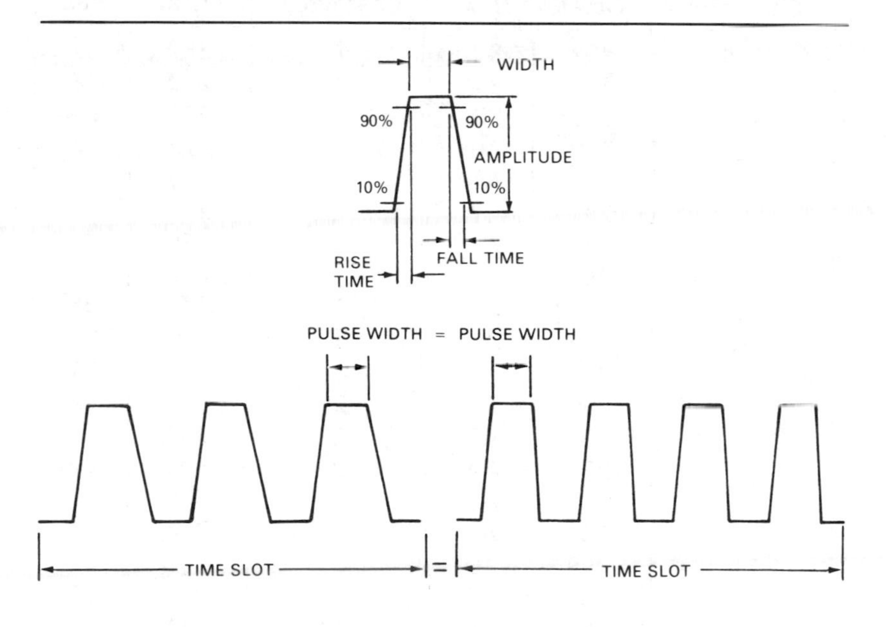

Fig. H-2. Effect of rise and fall times on maximum pulse rate. (Courtesy AMP Inc.)

When we use pulses to represent bits, we speak of speed as a bit rate, or so many bits per second (bits/s). The rising and falling of a pulse is akin to the rising and falling of a sine wave. Bit rate is, in a loose sense, a frequency. Bit rate and sine-wave frequency are related, but they are not the same. As a general rule, a frequency is 10 times the bit rate. Thus if a fiber-optic link is specified to have a maximum operating frequency of 200 MHz, its maximum bit rate is 20 Mbits/s. The term frequency, then, is used to include both sine-wave frequencies and bit rates.

Code Tables

Two codes of interest to readers of this book are the ASCII (American Standard Code for Information Interchange) code and the International Morse code. These are listed in Tables H-1 and H-2, respectively.

Table H-1. ASCII Keyboard Code

Character	Octal	Hex	Character	Octal	Hex
Bell	207	87	7	267	B7
LF	212	8A	8	270	B8
CR	215	8D	9	271	B9
Space	240	A0	:	272	BA
!	241	A1	;	273	BB
"	242	A2	<	274	BC
#	243	A3	=	275	BD
$	244	A4	>	276	BE
%	245	A5	?	277	BF
&	246	A6	@	300	C0
'	247	A7	A	301	C1
(	250	A8	B	302	C2
)	251	A9	C	303	C3
*	252	AA	D	304	C4
+	253	AB	E	305	C5
,	254	AC	F	306	C6
—	255	AD	G	307	C7
.	256	AE	H	310	C8
/	257	AF	I	311	C9
0	260	B0	J	312	CA
1	261	B1	K	313	CB
2	262	B2	L	314	CC
3	263	B3	M	315	CD
4	264	B4	N	316	CE
5	265	B5	O	317	CF
6	266	B6	P	320	D0

Table H-1. ASCII Keyboard Code (Continued)

Character	Octal	Hex	Character	Octal	Hex
Q	321	D1	Y	331	D9
R	322	D2	Z	332	DA
S	323	D3	[	333	DB
T	324	D4	\	334	DC
U	325	D5	]	335	DD
V	326	D6	∧	336	DE
W	327	D7	—	337	DF
X	330	D8			

NOTES:

1. LF = line feed, and CR = carriage return.
2. Codes are nine bits for octal (01 plus seven code bits) and eight bits for hexadecimal (1 plus seven code bits).

Table H-2. International Morse Code

Character	Code	Character	Code
A	·-	U	··-
B	-···	V	···-
C	-·-·	W	·--
D	-··	X	-··-
E	·	Y	-·--
F	··-·	Z	--··
G	--·	1	·----
H	····	2	··---
I	··	3	···--
J	·---	4	····-
K	-·-	5	·····
L	·-··	6	-····
M	--	7	--···
N	-·	8	---··
O	---	9	----·
P	·--·	0	-----
Q	--·-	,	--··--
R	·-·	.	·-·-·-
S	···	?	··--··
T	-		

Appendix I: Advantages of Fiber Optics

There are at least seven advantages to be gained through the use of fiber optics as an information-carrying medium:

- Greater bandwidth
- Smaller size and lighter weight
- Less signal loss
- Freedom from electromagnetic interference
- Confinement of signals
- Enhanced safety
- Lower overall cost

Bandwidth

The amount of information that can be carried through a system of any kind is limited by the frequency of the carrier wave. It is possible to transmit a certain amount of information on a carrier frequency of 1 kHz. Since the rule of thumb for data transmission is that the carrier frequency must be at least double the information frequency, a 1-kHz system could carry 500 Hz adequately. This would not accommodate much of the human voice, which reaches into the 8-kHz range, although it might serve to communicate by International Morse code.

Telephone companies usually limit the voice-carrying range of their equipment to just about 3 kHz for ordinary circuits, although their television cables handle several megahertz. Amplitude-modulated radio broadcasts usually have been limited to 10 kHz, whereas frequency-modulated broadcasts usually include the entire music range. Television transmissions in color require a bandwidth of about 4.5 MHz.

Coaxial-cable systems made possible the simultaneous transmission of several channels of television, by using a carrier wave of very high frequency. There is a limit to the frequency that a coaxial cable can carry with sufficient power to serve a large community of users. This limitation is a function of power versus the physical dimensions of the cable. As higher powers are attempted, the cable begins to arc across its dielectric.

Fiber optics handles the higher carrier frequencies, making possible not just a few simultaneous video channels, but hundreds of channels.

Size and Weight

While not every copper wire in an aircraft can be replaced with an optical fiber, collectively those which can be replaced by fibers weighing a few pounds may well weigh tons. Every pound that can be eliminated from a spacecraft is a pound that can be given over to payload. Transporting a huge reel of fiber cable by highway truck is an easy task compared to loading and carrying the many reels of copper cable that might do a roughly equivalent job of providing communications. As for size, conduits for telephone lines have already become a severe problem in large cities. Substitution of fiber optics for copper wires would permit an orders-of-magnitude increase in services with no increase in conduit capacity.

Signal Loss

Attenuation (signal loss per kilometer) is much less in a fiber-optics system than in an equivalent coaxial-cable system. Attenuation in a coaxial cable increases as the frequency of the signal increases, which is not the case with an optical fiber. This translates to greater distance between relay stations for the fiber system, and thus less overall cost.

Electromagnetic Interference

Stray magnetic fields such as those generated by power lines can create interference in sensitive equipment. Long telemetry leads can pick up considerable electromagnetically induced noise. Further, atomic reactions (such as atomic blasts at any altitude) have a devastating effect on ordinary wire and cable lines. Fiber lines are free from such interference. Fiber is a dielectric material and is thus free from electromagnetic effects.

Confinement

Wires carrying current radiate energy, which can generate interference in nearby circuits. Expensive shielding is used to prevent this in ordinary systems. Fibers do not radiate energy. They generate no interference to other systems, nor within the system they are intended to serve. Thus, the expense of shielding may be eliminated.

Safety

Short circuits, leaks, and hazards encountered in wiring systems can start fires, cause explosions, and even be a hazard to life. There is no spark when a light-carrying fiber is severed, nor can a fiber be "short circuited," in the electrical meaning of the term. A fiber conducting appreciable energy can be broken in an explosive environment quite safely.

Lower Cost

Although the cost of a one-meter length of high-quality fiber is generally higher than the cost of an identical length of copper wire, this is like trying to compare apples and bananas. The amount of data that can be transferred from one point to another via the fiber can be the equivalent in communications value to 900 strands of wire. Then there are additional advantages that tend to justify the higher first cost, such as the increase in security and the elimination of cross talk (when a bundle of wires is necessary to carry the same amount of information the single fiber can carry). The future of the fiber industry indicates that the cost of fibers will be reduced as manufacturing techniques are perfected. Copper wire is more likely to increase in cost.

In calculating the overall costs of a fiber system versus a conventional wired system, the reduced handling costs brought about by the smaller size and lighter weight of the fiber should be considered, as should the elimination of expensive shielding, the reduced number of repeaters, and the reduced likelihood of having to repair or replace a building damaged by fire or explosion. The advantages of fiber optics will often more than justify the added costs.

Appendix J: Glossary

Acceptance angle The angle within which a fiber will accept light for transmission along its core. The angle is measured from the centerline of the core.

Buffer Material for fiber covering without optical function.

Bundle A group of single fibers carried in a common covering.

Cable A single fiber or a bundle, including strengthening strands of opaque material if used, covered by a protective jacket.

Characteristic angle The angle at which a ray of light propagates along the fiber core.

Cladding A material that has a lower index of refraction than the light-carrying fiber and that sheathes the fiber core both for protection and to provide optical refraction or reflection.

Core The light-carrying material of a fiber, with a higher index of refraction than the cladding.

Coupling or coupler A device for optically connecting one fiber with another, or for connecting a fiber with a termination into or from a source or receiver.

Cross talk A leakage of communication energy from one conductor to another, principally in wire bundles. It can occur in fiber optics, and when measurable it is treated as leakage from one optical conductor to another.

Dispersion The spreading of a light pulse as it traverses a fiber; scattering of light rays.

Fiber The optical waveguide, or light-carrying core or conductor. It may be made of glass or other materials not generally referred to as fibrous.

Graded-index fiber A fiber with an index of refraction that decreases radially from the centerline of the core.

Index of refraction Mathematical ratio comparing the velocity of light in a vacuum to the velocity of light in the medium under consideration.

Material dispersion Spreading of the light pulse or beam caused by foreign material in the composition of the core. It causes variable propagation delays for the wavelengths being carried.

Modal dispersion A spreading of the light ray or pulse due to length differences encountered in the optical path of a multimode fiber.

Multimode fiber A fiber designed to propagate more than one mode at a given wavelength.

Numerical aperture A numerical parameter expressing the degree of light acceptance of an optical fiber.

Pulse dispersion The tendency of a pulse to spread out (take more time to initiate and conclude) due to modal and or material dispersion factors. *See also* Dispersion.

Scan To prepare data serially for transmission. In a television camera tube, for example, an electron beam traverses the image to be transmitted, interpreting as electrical signals the dark and light portions it contacts.

Single-mode fiber A fiber propagating in but one mode.

Step-index fiber A fiber in which the index of refraction changes abruptly at the boundary between core and cladding.

Waveguide dispersion Spreading of a light pulse caused by differences in the critical dimensions along the length of a fiber.

Appendix K: Bibliography

Books

AMP Inc. 1979. *AMP: Introduction to Fiber Optics* (HB 5444). Harrisburg, PA

Augat, Inc. 1979. *Augat Fiberoptic Data Transmission Systems Instruction Manual*. Attleboro, MA

Bendiksen, Leonard, and Intrieri, Charles 1979. *Laboratory Development of Fiber Optic Links* (P211-78). Harrisburg, PA: AMP Inc.

Bowen, Terry 1976. *Terminating Fiber Optic Bundles* (P179-76). Harrisburg, PA: AMP Inc.

Bowen, Terry 1978. *Low Cost Connectors for Single Optical Fibers* (P199-8). Harrisburg, PA: AMP Inc.

Canoga Data Systems 1979. *Data Communication Product Highlights*. Canoga Park, CA

ITT Electro-Optical Products Div. 1980. *Glass-on-Glass Optical Fiber End Preparation* (Technical Note R-4)

Kapany, N. S. 1967. *Fiber Optics, Principles and Applications*. New York: Academic Press

Kleekamp, C. W., and Metcalf, B. D. 1979. *Fiber Optics for Tactical Communications* (Report by the Mitre Corp. for the Electronic Systems Div., Air Force Systems Command, US Air Force). Air Force No. ESD-TR-79-121

Lubars, Herb 1979. *Optical Fiber Cable Systems*. Woodbridge, NJ: General Cable Co.

Motorola Semiconductor Products Inc. *Basic Experimental Fiber Optic Systems*. Phoenix, AZ

RCA 1979 – 80. *Optical Communications Products* (OPT-115). Lancaster, PA

RCA Solid-State Div., Electro Optics and Devices 1978. *RCA Electro-Optics Handbook*. Lancaster, PA.

Schumacher, William L. 1977. *Design Considerations for Single Fiber Connectors* (P196-77). Harrisburg, PA: AMP Inc.

Schumacher, William L. 1977. *Transmission Efficiencies of Optical Connections* (P198-77). Harrisburg, PA: AMP Inc.

Schumacher, William L. 1979. *Packaging Semiconductors for Optimum Optical Coupling* (P231-80). Harrisburg, PA: AMP Inc.

Superintendent of Documents, US Government Printing Office. *Measurement of Far-Field and Near-Field Radiation Patterns from Optical Fibers* (TN 1032). Washington, DC

Articles in Journals and Periodicals

Bojsza, Walter J. 1981. Optics: A Technology at Hand. *Microwaves* Vol. 20, No. 8.

Gadzinski, Eric, ed. 1980. Fiber Optics: Harnessing Light for a Bright Future. *Circuits Manufacturing*, Feb.

Greenwall, R. A., et al 1980. The Navy's Job — Fiber Optics. *International Fiber Optics and Communications* Vol. 1, No. 4.

Krueger, Alton H. 1980. Applying Fiber Optics to Photoelectric Switches. *Control Engineering* Vol. 27, No. 8.

Lutes, George F. 1980. Fiber Optics Transmit Clock Signal More Reliably. *NASA Tech Brief*, Winter.

McGowan, Michael J. 1980. Fiber Optic Link Runs 5 km at 150 Mbps. *Control Engineering* Vol. 27, No. 8.

Miller, S. E., Marcatil, E. A. J., and Li, T. 1973. Research Toward Optical Fiber Transmission Systems. *Proceedings of the IEEE* 61: 1703-1751.

Rainwater, J. Hank, and McMillan, R. W. 1980. A Hybrid Technology for Near-Millimeter Waves. *Microwaves* June: 76.

Ramsey, M. M., Hockham, K. A., and Kao, K. C. 1975. Propagation in Optical Fiber Waveguides. *Electrical Communication* Vol. 50, No. 5: 162.

Reprints

Cahners Publishing Co. Designer's Guide to Fiber Optics (A compilation of five articles from *EDN* magazine). Boston.

Electro-Optical Systems Design 1977. Backscatter Measurements on Optical Fibers (TN 1034). (Reprint from *Electro-Optical Systems Design*, Nov. 1977) Washington, DC: Superintendent of Documents, US Government Printing Office.

Keeler, Pete 1978. Fiber Optics: Alignment is the Fiber-Optic Connector's Main Job (Reprint from *Electronic Design* Vol. 26, No. 22). Attleboro, MA: Augat Interconnection Products Div.

Math, Irwin 1977. Basic Optical Data Links (Reprint from *Electro-Optical Systems Design*, Sept. 1977). Great Neck, NY: Math Associates, Inc.

Miller, Eric 1980. Introduction to Practical Fiber Optics (Reprint from *International Fiber Optics and Communications* Vol. 1, No. 5). National Semiconductor Corp.

Wey, Robert A. 1980. Fiberoptic Communications (Reprint from *Laser Focus Buyers Guide*, January 1980). Newton, MA: Advanced Technology Publications.

Periodicals

International Fiber Optics and Communications. Bimonthly. 167 Corey Rd., Brookline, MA 02146

Laser Focus with Fiberoptic Communications. Monthly. 1001 Watertown St., Newton, MA 02165

Index

READER SERVICE CARD

To better serve you, the reader, please take a moment to fill out this card, or a copy of it, for us. Not only will you be kept up to date on the Blacksburg Series books, but as an extra bonus, **we will randomly select five cards every month, from all of the cards sent to us during the previous month. The names that are drawn will win, absolutely free, a book from the Blacksburg Continuing Education Series.** Therefore, make sure to indicate your choice in the space provided below. For a complete listing of all the books to choose from, refer to the inside front cover of this book. Please, one card per person. Give everyone a chance.

In order to find out who has won a book in your area, call (703) 953-1861 anytime during the night or weekend. When you do call, an answering machine will let you know the monthly winners. Too good to be true? Just give us a call. Good luck.

If I win, please send me a copy of:

____

I understand that this book will be sent to me absolutely free, if my card is selected.

For our information, how about telling us a little about yourself. We are interested in your occupation, how and where you normally purchase books and the books that you would like to see in the Blacksburg Series. We are also interested in finding authors for the series, so if you have a book idea, write to The Blacksburg Group, Inc., P.O. Box 242, Blacksburg, VA 24060 and ask for an Author Packet. We are also interested in TRS-80, APPLE, OSI and PET BASIC programs.

My occupation is ______________________________
I buy books through/from ______________________________
Would you buy books through the mail? ______________________________
I'd like to see a book about ______________________________
Name ______________________________
Address ______________________________
City ______________________________
State ______________________________ Zip ______________

MAIL TO: BOOKS, BOX 715, BLACKSBURG, VA 24060
!!!!!PLEASE PRINT!!!!!